T0321589

Introduction to Electricity and Magnetism

Solutions to Problems

Introduction to Electricity and Magnetism

Solutions to Problems

John Dirk Walecka

College of William and Mary, USA

World Scientific

NEW JERSEY • LONDON • SINGAPORE • BEIJING • SHANGHAI • HONG KONG • TAIPEI • CHENNAI • TOKYO

Published by

World Scientific Publishing Co. Pte. Ltd.
5 Toh Tuck Link, Singapore 596224
USA office: 27 Warren Street, Suite 401-402, Hackensack, NJ 07601
UK office: 57 Shelton Street, Covent Garden, London WC2H 9HE

Library of Congress Cataloging-in-Publication Data
Names: Walecka, John Dirk, 1932– author.
Title: Introduction to electricity and magnetism : solutions to problems /
John Dirk Walecka (College of William and Mary, USA).
Description: Singapore ; Hackensack, NJ : World Scientific Publishing Co. Pte. Ltd., [2019] |
Includes bibliographical references and index.
Identifiers: LCCN 2019007559| ISBN 9789811202636 (pbk. ; alk. paper) |
ISBN 981120263X (pbk. ; alk. paper)
Subjects: LCSH: Electricity--Problems, exercises, etc. | Magnetism--Problems, exercises, etc.
Classification: LCC QC522 .W355 2019 | DDC 537--dc23
LC record available at https://lccn.loc.gov/2019007559

British Library Cataloguing-in-Publication Data
A catalogue record for this book is available from the British Library.

For any available supplementary material, please visit
https://www.worldscientific.com/worldscibooks/10.1142/11342#t=suppl

Printed in Singapore

For John and Ann

Preface

The present author has published several physics textbooks, either alone or in collaboration (see the *Bibliography*). All of these books are based on courses taught at either Stanford, where the author was on the faculty from 1960-1986, or at the College of William and Mary, where he served from 1992-2003.

There is one other course that I taught at Stanford in the spring of 1986 for which I have a good set of lecture notes. This was the calculus-based freshman physics course, Physics 53, on electricity and magnetism. It was a big course with over 400 students, taught in two lecture sections, with additional problem sessions taught by many good graduate teaching assistants, overseen by an excellent head TA, Janet Tate.[1] It was one of the most enjoyable courses I ever taught, and what made it such fun was the lecture demonstrations prepared by that master, Kurt Machein. I would go in each night before a lecture and practice the demonstrations for the next day under Kurt's watchful eye. They always went well and really enhanced and solidified the material.

There are many good introductory and more advanced electricity and magnetism texts. The one I used was [Ohanian (1985)], but see also, for example, [Halliday and Resnick (2013); Freedman *et al.* (2013); Purcell and Morin (2013); Griffiths (2017); Slater and Frank (2011); Schwarz (1987); Abraham and Becker (1949); Stratton (2008); Panofsky and Phillips (2005); Jackson (2009)], *etc.* The existence of all of these texts, and the inability to include the wonderful demonstrations, made me very reluctant to consider publishing those lectures.

On the other hand, those lectures *do* provide what in my opinion is a

[1]Physics 54 was an optional one-credit lab, graded P/NC, that went along with the course.

clear, self-contained, calculus-based introduction to a subject that together with classical mechanics, quantum mechanics, and modern physics lies at the heart of today's physics curriculum. A good introduction, even at the cost of some repetition, does allow one to approach the more advanced texts and monographs with better understanding and a deeper sense of appreciation. Furthermore, those lectures, although relatively concise, do take one from Coulomb's law to Maxwell's equations and special relativity in what to me is a lucid and logical fashion. The principles of electromagnetism have such an astonishing range of applicability. So, to round out the set of physics texts, and for my own enjoyment, I proceeded to convert those lectures into the book [Walecka (2018)].

That book contains an extensive set of accessible problems that enhances and extends the coverage. As an aid to teaching and learning, the present book provides the solutions to those problems. I hope that in using these texts, students and teachers alike can share some of the pleasure I took in writing them.

I would, once again, like to thank my editor Ms. Lakshmi Narayanan for her help and support on this project.

Williamsburg, Virginia
September 9, 2018

John Dirk Walecka
Governor's Distinguished CEBAF
Professor of Physics, Emeritus
College of William and Mary

Contents

PART 1

Electricity

Chapter 1

Introduction

There are no problems associated with the Introduction.

Chapter 2

Coulomb's Law

Problem 2.1 What is the gravitational force between the two charges in the example in section 2.2? Compare it to the electrostatic force.

Solution to Problem 2.1

The expression for the gravitational force between two masses (m_1, m_2) separated by $\vec{r} = \vec{r}_1 - \vec{r}_2$ is due to Newton

$$\vec{F}_{21} = -G\frac{m_1 m_2}{r^2}\,\hat{r} \qquad ;\ \text{gravitational force}$$

Here $\vec{F}_{21}$ is the force exerted on particle 1 by particle 2, and $\hat{r}$ is the unit vector $\hat{r} \equiv \vec{r}/r$. G is the gravitational constant

$$G = 6.67 \times 10^{-11}\,\frac{\text{Nm}^2}{\text{kg}^2}$$

We work in SI units where the force is measured in "newtons"

$$1\,\text{N} = 1\,\text{newton} = 1\frac{\text{kg-m}}{\text{sec}^2} = \frac{1}{4.45}\,\text{pound}$$

We are asked to compute the magnitude of the gravitational attraction of a completely ionized mole (12 grams) of ${}^{12}_{6}\text{C}$ separated from San Francisco to New York

$$r \approx 5000\,\text{km}$$

The number of atoms in one mole is given by Avogadro's number

$$N_{\text{A}} \equiv \text{Avogadro's number} = 6.02 \times 10^{23}\,/\text{mole}$$

The masses of the carbon nuclei and the electrons (six/atom) are

$$\begin{aligned} M_{^{12}\mathrm{C}} &\approx 12\,\mathrm{gm} \\ M_{6\mathrm{e}} &= 6N_{\mathrm{A}}m_e = 6 \times 6.02 \times 10^{23} \times 0.911 \times 10^{-27}\,\mathrm{gm} \\ &= 3.29 \times 10^{-3}\,\mathrm{gm} \end{aligned}$$

Hence

$$\begin{aligned} |F_{\mathrm{grav}}| &\approx 6.67 \times 10^{-11}\,\frac{\mathrm{Nm}^2}{\mathrm{kg}^2} \times \frac{(.012\,\mathrm{kg}) \times (3.29 \times 10^{-6}\,\mathrm{kg})}{(5 \times 10^6\,\mathrm{m})^2} \times \frac{1\,\mathrm{lb}}{4.45\,\mathrm{N}} \\ &= 2.37 \times 10^{-32}\,\mathrm{lb} \end{aligned}$$

This is miniscule compared with the magnitude of the corresponding electrostatic force as calculated in the text

$$|F_{\mathrm{coul}}| = 1.35 \times 10^4\,\mathrm{ton}$$

The electrostatic force is as large as it is because of the many-many oppositely charged particles packed into the solid.

Chapter 3

The Electric Field

Problem 3.1 Show that Eq. (3.9) still holds to the given order when the origin of the vector $\vec{r}$ lies at the midpoint of the displacement $\vec{d}$.[1]

Solution to Problem 3.1

Suppose the origin of the field vector $\vec{r}$ lies at the midpoint of the displacement $\vec{d}$ in the dipole (Fig. 3.1 below). The field is then given by

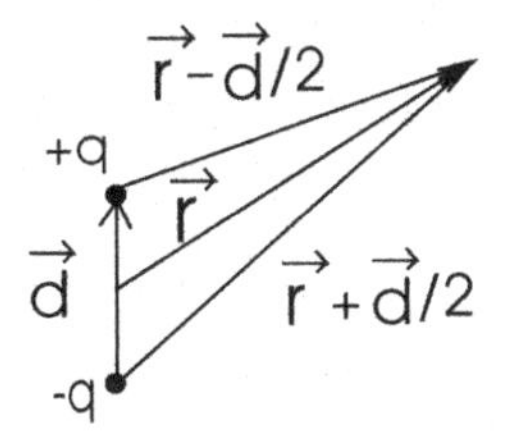

Fig. 3.1 Origin of the vector $\vec{r}$ lies at the midpoint of the dipole displacement $\vec{d}$.

$$\vec{E}(\vec{r}) = \frac{q}{4\pi\varepsilon_0}\left[\frac{(\vec{r}-\vec{d}/2)}{|(\vec{r}-\vec{d}/2)|^3} - \frac{(\vec{r}+\vec{d}/2)}{|(\vec{r}+\vec{d}/2)|^3}\right]$$

Write

$$\frac{1}{|(\vec{r}-\vec{d}/2)|^3} = [(\vec{r}-\vec{d}/2)^2]^{-3/2} = [r^2 - \vec{r}\cdot\vec{d} + d^2/4]^{-3/2}$$

Then retaining terms up through order d, one has from the binomial theo-

[1] *Hint:* Start from a new figure.

rem in Eq. (3.6)

$$\frac{1}{|(\vec{r}-\vec{d}/2)|^3} \approx \frac{1}{r^3}\left[1+\frac{3}{2}\frac{\vec{r}\cdot\vec{d}}{r^2}\right]$$

With the use of this expression

$$\vec{E} \approx \frac{q}{4\pi\varepsilon_0 r^3}\left[\left(\vec{r}-\frac{\vec{d}}{2}\right)\left(1+\frac{3}{2}\frac{\vec{r}\cdot\vec{d}}{r^2}\right)-\left(\vec{r}+\frac{\vec{d}}{2}\right)\left(1-\frac{3}{2}\frac{\vec{r}\cdot\vec{d}}{r^2}\right)\right]$$

Hence, as requested, we see that the field is still given through order d/r by the result in Eq. (3.9)

$$\vec{E}(\vec{r}) = \frac{q}{4\pi\varepsilon_0}\frac{1}{r^3}\left[3\hat{r}(\hat{r}\cdot\vec{d}\,)-\vec{d}\,\right] \qquad ; \text{ dipole field}$$
$$; \ d/r \ll 1$$

Problem 3.2 Make a good numerical calculation of the dipole field in Eq. (3.9).

Solution to Problem 3.2

We first measure all distances in units of the dipole distance d, and write the dipole field expression as

$$\vec{E}(\vec{r}) = \frac{q}{4\pi\varepsilon_0\, d^2}\frac{1}{(r/d)^3}\left[3\hat{r}(\hat{r}\cdot\hat{d}\,)-\hat{d}\,\right] \qquad ; \text{ dipole field}$$
$$; \ d/r \ll 1$$

For ease of plotting, we then convert to cartesian coordinates, where we put the dipole along the z-axis with $\hat{d}=\hat{z}$, a unit vector in the z-direction. Then

$$\left(\frac{r}{d}\right)^2 = x^2+z^2 \qquad ; \ \hat{r} = \frac{x}{(x^2+z^2)^{1/2}}\,\hat{x}+\frac{z}{(x^2+z^2)^{1/2}}\,\hat{z}$$

To construct the lines of force, we need the unit vector $\hat{E}(\vec{r})$. The components of $\vec{E}$ at each point are given by

$$E_z = \frac{q}{4\pi\varepsilon_0 d^2}\frac{1}{(r/d)^3}\left[\frac{3z^2}{x^2+z^2}-1\right]$$
$$E_x = \frac{q}{4\pi\varepsilon_0 d^2}\frac{1}{(r/d)^3}\left[\frac{3xz}{x^2+z^2}\right]$$

The length of the vector $\vec{E}$ is[2]

$$|\vec{E}| = (E_x^2 + E_z^2)^{1/2} = \frac{q}{4\pi\varepsilon_0\, d^2}\frac{1}{(r/d)^3}\left[\frac{x^2+4z^2}{x^2+z^2}\right]^{1/2}$$

The components of the unit vector $\hat{E}(\vec{r})$ are then given by[3]

$$[\hat{E}(\vec{r})]_x = \frac{E_x}{|\vec{E}|} = \frac{3xz}{[(x^2+z^2)(x^2+4z^2)]^{1/2}}$$

$$[\hat{E}(\vec{r})]_z = \frac{E_z}{|\vec{E}|} = \frac{2z^2-x^2}{[(x^2+z^2)(x^2+4z^2)]^{1/2}}$$

The unit vector $\hat{E}$ then gives the tangent to the line of force

$$\hat{E}(\vec{r}) = [\hat{E}(\vec{r})]_x\, \hat{x} + [\hat{E}(\vec{r})]_z\, \hat{z} \qquad \text{; tangent unit vector}$$

The line of force is constructed by starting at some point (x_1, z_1) and stepping out in small steps dl along the tangent to the curve

$$x_{i+1} = x_i + dl\left[\hat{E}(x_i.z_i)\right]_x$$

$$z_{i+1} = z_i + dl\left[\hat{E}(x_i.z_i)\right]_z$$

Some numerical results obtained with Mathcad 7 are shown in Fig. 3.2 below. While the sketch of the field lines in Fig. 3.6 in the text may not be bad when close to the dipole, the actual dipole field lines of force look quite different, and quite characteristic, far away.

Problem 3.3 A particle of mass and charge (m, q) starts from rest a distance z_0 above the sheet of charge in Fig. 3.7. Show that its velocity and position after time t are

$$v_z = \left(\frac{\sigma q}{2m\varepsilon_0}\right)t \qquad ;\ z = \frac{1}{2}\left(\frac{\sigma q}{2m\varepsilon_0}\right)t^2 + z_0$$

Solution to Problem 3.3

The electric field above the sheet of charge in Fig. 3.7 in the text is given in Eq. (3.16)

$$\vec{E} = \frac{\sigma}{2\varepsilon_0}\hat{z} \qquad \text{; sheet of charge in } (x,y)\text{-plane}$$

[2]Use $9x^2z^2 + (2z^2 - x^2)^2 = (x^2+z^2)(x^2+4z^2)$.

[3]See the solution to Prob. 11.3.

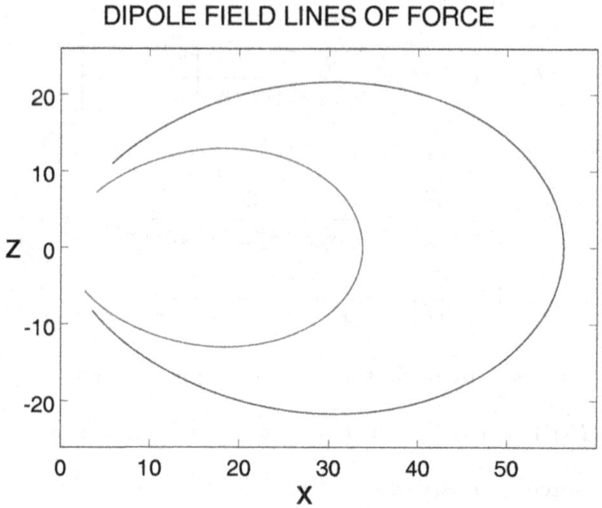

DIPOLE FIELD LINES OF FORCE

Z: 20, 10, 0, -10, -20

X: 0, 10, 20, 30, 40, 50

Fig. 3.2 Numerical calculation of the lines of force in the dipole field from Prob. 3.1. All distances are measured in units of the dipole length d, and there is a dipole $\vec{d} = \hat{z}$ at the origin. The analytic expression used holds for $(x^2 + z^2)^{1/2} \gg 1$. The field lines flow in the clockwise direction and have cylindrical symmetry about the z-axis.

The motion of a particle with mass and charge (m, q) in this field is then given by Newton's second law

$$m\frac{d\vec{v}}{dt} = \frac{\sigma q}{2\varepsilon_0}\hat{z}$$

Suppose the particle starts with $(z, v_z) = (z_0, 0)$ and moves in the z-direction with $\vec{v} = v_z\hat{z}$. The first integral of the above gives

$$v_z = \left(\frac{\sigma q}{2m\varepsilon_0}\right)t$$

Since $v_z = dz/dt$, a second integral then gives

$$z = \frac{1}{2}\left(\frac{\sigma q}{2m\varepsilon_0}\right)t^2 + z_0$$

Chapter 4

Gauss' Law

Problem 4.1 Suppose the positive point charge q lies *outside* the closed surface S in section 4.2. Show the integrated electric flux from that charge then *vanishes*.[1] Conclude that this charge does not contribute to Gauss' law.

Solution to Problem 4.1

Suppose the charge lies *outside* of the closed surface S, and for simplicity, let S be a simple convex surface. Draw a cone from the charge that is *tangent* to the convex surface (Fig. 4.1 below). In the tangent plane at the intersection of the cone with the surface, one has $\hat{r} \cdot d\vec{S} = 0$. Then in Eqs. (4.4), the only difference in the calculation of the electric flux for that part of the surface on the right with outward pointing normal, lying within the solid angle Ω_0, is that $\hat{r} \cdot d\vec{S}$ is *negative.*

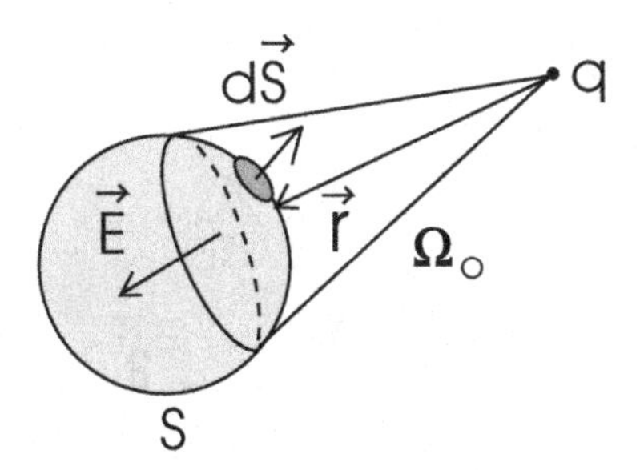

Fig. 4.1 Configuration for derivation of Gauss' law if q lies *outside* of S. The indicated cone is drawn tangent to the convex surface. In the tangent plane at the intersection of the cone with the surface, one has $\hat{r} \cdot d\vec{S} = 0$.

[1] *Hint:* Draw tangent cones and consider the solid angles.

For that part of the solid angle, Eq. (4.4) becomes

$$\vec{E} \cdot d\vec{S} = -|\vec{E}|r^2 d\Omega \qquad ; \text{ electric flux}$$

For the reminder of the surface, $\hat{r} \cdot d\vec{S}$ is *positive* and Eqs. (4.4) hold. Equations (4.6)–(4.7) then give

$$\begin{aligned}\int_S \vec{E} \cdot d\vec{S} &= -\int_{\Omega_0} \frac{q}{4\pi\varepsilon_0}\frac{1}{r^2}(r^2 d\Omega) + \int_{\Omega_0} \frac{q}{4\pi\varepsilon_0}\frac{1}{r^2}(r^2 d\Omega) \\ &= \frac{q}{4\pi\varepsilon_0}\left[-\int_{\Omega_0} d\Omega + \int_{\Omega_0} d\Omega\right] = 0 \qquad ; \text{ point charge}\end{aligned}$$

We conclude that if the positive point charge q lies *outside* the closed surface S in section 4.2, the integrated electric flux from that charge *vanishes.* Hence that charge does not contribute to Gauss' law.

Problem 4.2 Derive Eq. (4.15) by integrating Coulomb's law for a line of charge.

Solution to Problem 4.2

The net field must be perpendicular to the line of charge (see Fig. 4.2 below)

$$\vec{E} = E_\rho\, \hat{\rho}$$

The contribution from an element of charge $\lambda\, dz$ along the line at a distance

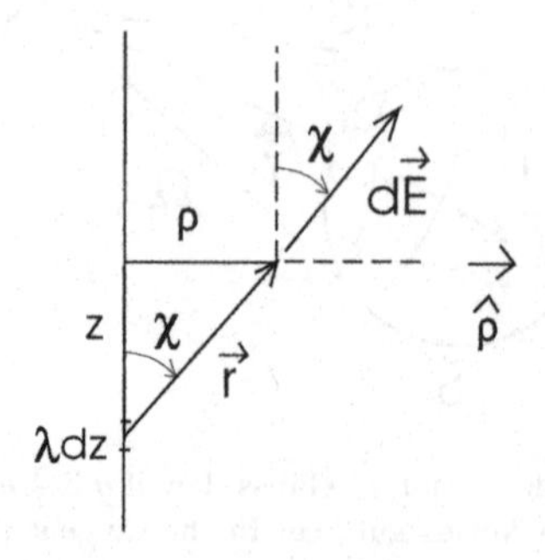

Fig. 4.2 Integration of Coulomb's law for a line of charge.

ρ from the line in the $\hat{\rho}$-direction is

$$dE_\rho = \frac{\lambda}{4\pi\varepsilon_0}\frac{dz}{(z^2+\rho^2)}\cos(\pi/2-\chi) = \frac{\lambda}{4\pi\varepsilon_0}\frac{dz}{(z^2+\rho^2)}\sin\chi$$
$$= \frac{\lambda}{4\pi\varepsilon_0}\frac{\rho\, dz}{(z^2+\rho^2)^{3/2}}$$

Introduce $t \equiv z/\rho$, and integrate along the line from $-\infty$ to ∞

$$E_\rho = \frac{\lambda}{4\pi\varepsilon_0}\frac{1}{\rho}\int_{-\infty}^{\infty}\frac{dt}{(1+t^2)^{3/2}}$$

From Mathcad 7, the integral is

$$\int_{-\infty}^{\infty}\frac{dt}{(1+t^2)^{3/2}} = 2$$

Hence we derive Eq. (4.15) directly from Coulomb's law

$$E_\rho = \frac{\lambda}{2\pi\varepsilon_0}\frac{1}{\rho}$$

Problem 4.3 Start from Fig. 3.8(b), and use Gauss' law to derive the field in Eq. (3.17).

Solution to Problem 4.3

Figure. 3.8(b)[2] exhibits the electric field between two charged plates

$$\vec{E} = E_z\,\hat{z}$$

Here the electric field lines start on the positive charges and end on the negatives charges; they lie *between* the plates.

Draw the gaussian pillbox of Fig. 4.4 that surrounds an element of area A on the positively charged plate. When Gauss' law is applied to the pillbox, the only non-zero contribution arises from the surface *between* the plates

$$\int_S \vec{E}\cdot d\vec{S} = E_z A + 0 + 0 = \frac{\sigma A}{\varepsilon_0}$$

where σ is the surface charge density (charge/area). This reproduces the expression for the field in Eq. (3.17)

$$\vec{E} = \frac{\sigma}{\varepsilon_0}\hat{z} \qquad ; \text{ between two sheets with opposite charge}$$

[2]Figure numbers now refer to the text, unless explicitly noted.

Problem 4.4 Consider a metallic conductor with a hole inside of it. Start with a solid conductor where the field vanishes inside and the charge is all on the surface (see Fig. 4.6). Now take a chunk out of its interior. Convince yourself that nothing changes. Hence, conclude that the field $\vec{E}$ still vanishes inside a hole in the conductor.

Solution to Problem 4.4

Consider a solid, charged conductor. There is no electric field *inside* the conductor, and the charge all *sits on the surface.* The actual charge arrangement on the surface may be very complicated, but it is arranged so that the field *vanishes* inside. Now remove a small amount of the (neutral) conductor from inside. There is no charge there, and no field, so there are no active electrical forces there. The charge on the surface is undisturbed, as there are no new electrical forces acting on the surface, and thus the charge on the surface continues to be arranged to produce a vanishing electric field everywhere inside the conductor, and hence, in the hole. Now simply enlarge the hole!

Problem 4.5 Show that Eqs. (5.56) and (5.57) reproduce Coulomb's law for the collection of charges in Fig. 5.6.[3]

Solution to Problem 4.5

Equations (5.18) and (5.57) read

$$V(\vec{r}) = \frac{1}{4\pi\varepsilon_0}\sum_{i=1}^{N}\frac{q_i}{|\vec{r}-\vec{r}_i|} \qquad ; \text{ electrostatic potential}$$

$$\vec{E}(\vec{r}) = -\vec{\nabla}V(\vec{r})$$

Use the *hint* to write each denominator as

$$\sqrt{(\vec{r}-\vec{r_i})^2} = \left[(x-x_{0i})^2 + (y-y_{0i})^2 + (z-z_{0i})^2\right]^{1/2}$$

Now repeat the analysis of the derivatives in the gradient in section 5.9.2 to obtain the field contribution from each charge as

$$\vec{E}_i(\vec{r}) = \frac{q}{4\pi\varepsilon_0}\frac{(\vec{r}-\vec{r}_i)}{[(\vec{r}-\vec{r}_i)^2]^{3/2}}$$

[3] *Hint:* Write $(\vec{r}-\vec{r_i})^2 = (x-x_{0i})^2 + (y-y_{0i})^2 + (z-z_{0i})^2$. Note Eq. (5.56) should be identical to Eq. (5.18).

where

$$\vec{r} - \vec{r_i} = (x - x_{0i})\,\hat{x} + (y - y_{0i})\,\hat{y} + (z - z_{0i})\,\hat{z}$$

This reproduces Coulomb's law for the collection of charges

$$\vec{E}(\vec{r}) = \sum_{i=1}^{N} \vec{E}_i(\vec{r})$$

Problem 4.6 An infinite line of positive charge density λ per unit length is surrounded by a metallic conducting cylinder, of inner radius a and outer radius b, whose axis lies along the line of charge. The cylinder is also of infinite length. Let $\vec{\rho}$ be a radius vector perpendicular to the line of charge.

(a) Show that the electric field for $0 < \rho < a$ is given by

$$\vec{E} = \frac{\lambda}{2\pi\varepsilon_0}\frac{\hat{\rho}}{\rho}$$

(b) What is the electric field for $a < \rho < b$?
(c) What is the electric field for $\rho > b$?

Solution to Problem 4.6

By symmetry, the field here points in the radial direction

$$\vec{E} = E_\rho\,\hat{\rho}$$

(a) Gauss' law applied to a concentric gaussian cylinder of height l and radius $\rho < a$, contained *inside of* the metallic conducting cylinder, gives

$$2\pi\rho l E_\rho = \frac{1}{\varepsilon_0}\lambda l$$

Hence

$$\vec{E} = \frac{\lambda}{2\pi\varepsilon_0}\frac{\hat{\rho}}{\rho} \qquad ; \rho < a$$

(b) The electric field must vanish in the *interior* of the surrounding metallic conducting cylinder

$$\vec{E} = 0 \qquad ; a < \rho < b$$

By Gauss' law, this implies a corresponding cancelling charge $-\lambda l$ has been induced on the inside surface of the surrounding metallic conducting cylinder.

(c) Since charge is conserved, an opposite charge λl must be induced on the *outside* of the neutral metallic conducting cylinder. Gauss' law applied to the gaussian cylinder with an expanded radius $\rho > b$ then reproduces the result in part (a)

$$\vec{E} = \frac{\lambda}{2\pi\varepsilon_0}\frac{\hat{\rho}}{\rho} \qquad ; \rho > b$$

The metallic cylinder simply serves as a *shield* to eliminate the electric field inside of it.

Problem 4.7 More complicated electric field configurations can be obtained by combining solvable components using the principle of *superposition*. For example, suppose there are three sheets of charge, with surface charge density σ, in the (x, y), (x, z), and (y, z)-planes. Show the field in the first octant is

$$\vec{E} = \frac{\sigma}{2\varepsilon_0}(\hat{x} + \hat{y} + \hat{z})$$

Solution to Problem 4.7

It is assumed that the three sheets of charge do not affect each other.

Consider the right-handed cartesian coordinate system in Fig. 11.1. The field in the $\hat{z}$ direction above the (x, y)-plane is

$$\vec{E}_1 = \frac{\sigma}{2\varepsilon_0}\hat{z}$$

Similarly, the field in the $\hat{x}$ direction in front of the (y, z)-plane is

$$\vec{E}_2 = \frac{\sigma}{2\varepsilon_0}\hat{x}$$

The field in the $\hat{y}$ direction to the right of the (x, z)-plane is

$$\vec{E}_3 = \frac{\sigma}{2\varepsilon_0}\hat{y}$$

Hence, by *superposition*, the field in the first octant is

$$\vec{E} = \vec{E}_1 + \vec{E}_2 + \vec{E}_3 = \frac{\sigma}{2\varepsilon_0}(\hat{x} + \hat{y} + \hat{z})$$

Chapter 5

The Electrostatic Potential

Problem 5.1 The unit of potential, the *volt*, is defined by

$$1\,\text{volt} \equiv \frac{1\,\text{J}}{1\,\text{C}} = \frac{1\,\text{Nm}}{1\,\text{C}} \equiv 1\,\text{V}$$

The *electron volt* is a unit of energy defined by

$$1\,\text{eV} \equiv |e| \times 1\,\text{volt}$$

Show[1]

$$1\,\text{eV} = 1.602 \times 10^{-19}\,\text{J}$$

Solution to Problem 5.1

The electron-volt (eV) is the energy an electron (or proton) gets when it is accelerated through a potential difference of one volt. Hence this energy, which is the product of the charge times the potential difference, is

$$1\,\text{eV} \equiv |e| \times 1\,\text{volt}$$

The magnitude of the charge on the electron is

$$|e| = 1.602 \times 10^{-19}\,\text{C}$$

In SI units

$$1\,\text{volt}\,(\text{V}) \times 1\,\text{coulomb}\,(\text{C}) = 1\,\text{joule}\,(\text{J})$$

Therefore

$$1\,\text{eV} = = 1.602\,\times 10^{-19}\,\text{VC} = 1.602 \times 10^{-19}\,\text{J}$$

[1]See appendix B in the text.

Problem 5.2 What is the electrostatic potential in the first octant in Prob. 4.7?

Solution to Problem 5.2

If the (x, y), (x, z), and (y, z)-planes are grounded in Prob. 4.7, and the charge sheets do not affect each other, then the electrostatic potential in the first octant is[2]

$$V(\vec{x}) = -\frac{\sigma}{2\varepsilon_0}(x + y + z)$$

The justification is that the gradient

$$\vec{\nabla} = \hat{x}\frac{\partial}{\partial x} + \hat{y}\frac{\partial}{\partial y} + \hat{z}\frac{\partial}{\partial z}$$

then gives the electric field through

$$\begin{aligned} \vec{E}(\vec{x}) &= -\vec{\nabla}V(\vec{x}) \\ &= \frac{\sigma}{2\varepsilon_0}(\hat{x} + \hat{y} + \hat{z}) \end{aligned}$$

Problem 5.3 The electrostatic potential above the sheet of charge in Fig. 3.7 is given in Eq. (5.22). Use energy conservation to derive the expression for the position in Prob. 3.3.

Solution to Problem 5.3

The electrostatic potential in Eq. (5.22) above the sheet of charge in Fig. 3.7 is

$$V(2) - V(1) = -\frac{\sigma}{2\varepsilon_0}(z_2 - z_1)$$

The potential decreases linearly as we move up away from the sheet.

The statement of energy conservation for the particle in Prob. 3.3 moving in the z-direction above the sheet is then

$$T(2) - T(1) = qV(1) - qV(2)$$

Hence

$$\frac{m}{2}v_z^2 = \frac{\sigma q}{2\varepsilon_0}(z - z_0) \qquad ; \text{ energy conservation}$$

[2]Any additional constant potential does not affect the argument.

Substitution of the results in Prob. 3.3 obtained with a first and second integral of Newton's second law verifies that the above expression indeed holds

$$\frac{m}{2}\left[\left(\frac{\sigma q}{2m\varepsilon_0}\right)^2 t^2\right] = \frac{\sigma q}{2\varepsilon_0}\left[\frac{1}{2}\left(\frac{\sigma q}{2m\varepsilon_0}\right)t^2\right]$$

However, energy conservation directly relates the position $z - z_0$ with the kinetic energy $mv_z^2/2$, without the necessity of explicitly finding that second integral.

Problem 5.4 With very many charges, the distribution of charges can be described by a continuous *charge density* $\rho(\vec{x}\,)$, which gives the total charge per unit volume at the position $\vec{x}$. Show that in this continuum limit, the electrostatic potential in Eq. (5.18) becomes

$$V(\vec{x}\,) = \frac{1}{4\pi\varepsilon_0}\int \frac{\rho(\vec{x}\,')d^3x'}{|\vec{x} - \vec{x}\,'|}$$

Solution to Problem 5.4

Equation (5.18) for the electrostatic potential of a collection of N charges reads

$$V(\vec{r}\,) = \frac{1}{4\pi\varepsilon_0}\sum_{i=1}^{N}\frac{q_i}{|\vec{r} - \vec{r}_i\,|} \qquad ; \text{ electrostatic potential}$$

Divide space up into many small cells, and let $\vec{r}_i$ denote the center of the ith cell. With very many charges, we can introduce the *charge density* $\rho(\vec{x}_i)$ where the number of charges in the ith cell is

$$\text{number of charges in } i\text{th cell} = \rho(\vec{x}_i)d^3r_i$$

The electrostatic potential then takes the form

$$V(\vec{r}\,) = \frac{1}{4\pi\varepsilon_0}\sum_{i}\frac{\rho(\vec{x}_i)d^3r_i}{|\vec{r} - \vec{r}_i\,|} \qquad ; \text{ electrostatic potential}$$

where the sum now goes over all the cells. In the limit where the cell size goes to zero, and the charge density becomes a continuous function, this expression becomes an *integral*

$$V(\vec{x}\,) = \frac{1}{4\pi\varepsilon_0}\int \frac{\rho(\vec{x}\,')d^3x'}{|\vec{x} - \vec{x}\,'|}$$

This is a very important result because it allows one to find the electrostatic potential of an *arbitrary charge distribution* in electrostatics, from which the electric field is obtained through the gradient

$$\vec{E}(\vec{x}\,) = -\vec{\nabla}V(\vec{x}\,)$$

Chapter 6

Electric Energy

Problem 6.1 A particle of mass m and charge q_{test} is placed halfway between two heavy, fixed charges $+q$, which are separated by a distance d.

(a) What is the electrostatic potential at the initial position of the particle?

(b) The particle is moved slightly from its initial position and released from rest; it accelerates away. Show that the velocity of the particle when it is very far away from its initial position is given by

$$v_\infty = \left[\frac{qq_{\text{test}}}{4\pi\varepsilon_0}\frac{8}{md}\right]^{1/2}$$

Solution to Problem 6.1

(a) The electrostatic potential at the initial position of the particle, with two heavy, fixed charges $+q$ a distance $d/2$ away, and with $V(\infty)$ as ground, is

$$V = V_1 + V_2 = \frac{q}{4\pi\varepsilon_0}\frac{1}{d/2} + \frac{q}{4\pi\varepsilon_0}\frac{1}{d/2} = \frac{q}{4\pi\varepsilon_0}\frac{4}{d}$$

(b) From energy conservation, when the particle has moved far away

$$\frac{1}{2}mv_\infty^2 = q_{\text{test}}V = \frac{qq_{\text{test}}}{4\pi\varepsilon_0}\frac{4}{d}$$

Hence

$$v_\infty = \left[\frac{qq_{\text{test}}}{4\pi\varepsilon_0}\frac{8}{md}\right]^{1/2}$$

Problem 6.2 Now suppose the two heavy charges $+q$ in Prob. 6.1 are no longer fixed, but are free to move. What is the sum of *their* kinetic energies when they are very far apart?

Solution to Problem 6.2

The initial electrostatic energy of the two heavy fixed charges is

$$U = \frac{q^2}{4\pi\varepsilon_0}\frac{1}{|\vec{r}_1 - \vec{r}_2|} = \frac{q^2}{4\pi\varepsilon_0}\frac{1}{d}$$

From energy conservation, when the particles are far apart, the sum of their kinetic energies $T_\infty = T_1 + T_2$ must be

$$T_\infty = U$$

Chapter 7

Capacitors and Dielectrics

Problem 7.1 A charge of 0.1 C is placed on a system of two 10 mF capacitors connected in parallel.[1] What is the voltage across the system? What is the voltage if they are connected in series?

Solution to Problem 7.1

For two capacitors in *parallel*, the equivalent total capacity is

$$\mathcal{C} = \mathcal{C}_1 + \mathcal{C}_2$$

Hence

$$\mathcal{C} = 10\,\text{mF} + 10\,\text{mF} = 20\,\text{mF} \qquad ; \text{ parallel}$$

The voltage across the system is then

$$\Delta V = \frac{Q}{\mathcal{C}} = \frac{0.1\,\text{C}}{0.2 \times 10^{-5}\,\text{F}} = 0.5 \times 10^5\,\text{V}$$

If the capacitors are in *series*, then

$$\frac{1}{\mathcal{C}} = \frac{1}{\mathcal{C}_1} + \frac{1}{\mathcal{C}_2}$$

Thus

$$\frac{1}{\mathcal{C}} = \frac{1}{10\,\text{mF}} + \frac{1}{10\,\text{mF}} = \frac{1}{5\,\text{mF}} \qquad ; \text{ series}$$

In this case

$$\Delta V = = \frac{0.1\,\text{C}}{0.5 \times 10^{-4}\,\text{F}} = 0.2 \times 10^4\,\text{V}$$

[1]Here mF means "millifarad", or 10^{-3}F.

Problem 7.2 Two concentric oppositely charged conducting cylinders have the region between them filled with a dielectric with dielectric constant κ.

(a) Use Gauss' law in the presence of dielectrics to determine the displacement field $\vec{D}$ and electric field $\vec{E}$ everywhere.[2]

(b) What is the capacity per unit length of the device?

Solution to Problem 7.2

(a) The configuration is shown in Fig. 7.1 below.

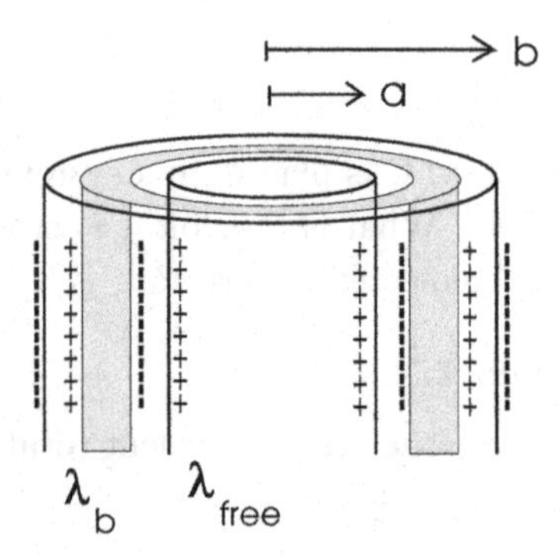

Fig. 7.1 Two co-axial ("concentric") oppositely charged conducting cylinders of radii a and b respectively have the region between them filled with a dielectric with dielectric constant κ. The charge per unit length on the conducting cylinders is $\lambda_{\rm free}$, and on the dielectric is λ_b. It is assumed the dielectric fills the region between the cylinders.

From symmetry, the electric field is radial $\vec{E} = E_\rho\hat{\rho}$. Draw a co-axial gaussian cylinder of height l and radius ρ *between* the charged conducting cylinders as in Prob. 4.6. Here there is a free charge per unit length $\lambda_{\rm free}$ on the cylinders and an induced charge per unit length λ_b on the dielectric, which we assume fills the region between the cylinders. The electric field $\vec{E}$ comes from the total charge inside the gaussian cylinder, and the displacement field $\vec{D}$ comes from the free charge. Gauss' law then says for $\vec{E}$

$$
\begin{aligned}
2\pi\rho l E_\rho &= \frac{1}{\varepsilon_0} l(\lambda_{\rm free} - \lambda_b) \\
\vec{E} &= \frac{(\lambda_{\rm free} - \lambda_b)}{2\pi\varepsilon_0}\frac{\hat{\rho}}{\rho} \qquad ; a < \rho < b
\end{aligned}
$$

[2]Recall section 4.3.2.

A similar calculation for $\vec{D}$ gives

$$\vec{D} = \frac{\lambda_{\rm free}}{2\pi\varepsilon_0}\frac{\hat{\rho}}{\rho} \qquad ; \; a < \rho < b$$

The fields in the dielectric are related by Eq. (7.32)[3]

$$\vec{D} = \kappa \vec{E} \qquad ; \text{ in dielectric}$$

where κ is the dielectric constant. Hence, the charges are also related by the dielectric constant

$$\lambda_{\rm free} - \lambda_b = \frac{1}{\kappa}\lambda_{\rm free}$$

just as in Eqs. (7.18).

(b) The potential difference between the plates is calculated exactly as in section 5.4.2

$$\Delta V = \frac{(\lambda_{\rm free} - \lambda_b)}{2\pi\varepsilon_0}\ln\left(\frac{b}{a}\right) = \frac{1}{\kappa}\frac{\lambda_{\rm free}}{2\pi\varepsilon_0}\ln\left(\frac{b}{a}\right)$$

The charge per unit length on the conducting cylinders is $\lambda_{\rm free}$. It follows that the capacity per unit length of the device $\mathcal{C}_l$ is

$$\mathcal{C}_l = \frac{\lambda_{\rm free}}{\Delta V} = \kappa \mathcal{C}_l^0 \qquad ; \; \mathcal{C}_l^0 \equiv \frac{2\pi\varepsilon_0}{\ln(b/a)}$$

Problem 7.3 Suppose the dielectric slab with dielectric constant κ only fills a fraction x of the distance d between the plates of the parallel-plate capacitor in Fig. 7.6. What is the capacity of the device?

Solution to Problem 7.3

The electric field inside the condenser, and corresponding voltage difference across it, are now modified. The analysis in sections 7.4–7.5 gives for the electric field

$$\begin{aligned}\vec{E} &= \frac{(\sigma_{\rm free} - \sigma_b)}{\varepsilon_0}\hat{z} = \frac{\sigma_{\rm free}}{\kappa\varepsilon_0}\hat{z} & ; \; 0 < z < xd \\ &= \frac{\sigma_{\rm free}}{\varepsilon_0}\hat{z} & ; \; xd < z < d\end{aligned}$$

The corresponding voltage difference across the plates is

$$\Delta V = \left[\frac{1}{\kappa}x + (1 - x)\right]\frac{\sigma_{\rm free}d}{\varepsilon_0}$$

[3]Compare Prob. 7.4.

Hence the capacity is

$$\mathcal{C} = \frac{\sigma_{\text{free}} A}{\Delta V} = \left[\frac{\kappa}{x + \kappa(1-x)}\right] \mathcal{C}_0 \qquad ; \; \mathcal{C}_0 = \frac{\varepsilon_0 A}{d}$$

This has the proper limits:

- For $x = 0$ (no dielectric), $\mathcal{C} = \mathcal{C}_0$;
- For $x = 1$ (dielectric fills capacitor), $\mathcal{C} = \kappa \mathcal{C}_0$.

Problem 7.4 The properties of a dielectric can be described with a *polarization vector* $\vec{P}$ defined so that *in* the dielectric

$$\vec{D} = \vec{E} + \vec{P} \qquad ; \text{ polarization}$$
$$\vec{P} = \chi_e \vec{E} \qquad ; \text{ electric susceptibility}$$

(a) Refer to Fig. 7.6. Show that for the dielectric slab in the condenser, the polarization gives the bound surface charge density

$$\varepsilon_0 \vec{P} = \sigma_b \, \hat{z}$$

(b) Show that for the slab, $\varepsilon_0 \vec{P}$ is the dipole moment per unit volume.

Solution to Problem 7.4

(a) From Eq. (7.16) in the text[4]

$$\vec{E} = \vec{D} - \vec{P} = \frac{1}{\varepsilon_0}(\sigma_{\text{free}} - \sigma_b)\, \hat{z}$$

Hence

$$\varepsilon_0 \vec{D} = \sigma_{\text{free}} \, \hat{z} \qquad ; \; \varepsilon_0 \vec{P} = \sigma_b \, \hat{z}$$

(b) The total dipole moment of the slab in Fig. 7.6 is $Q_b \, \vec{d} = \sigma_b A(\hat{z} d)$, and the volume of the slab is $v = Ad$. Thus the dipole moment per unit volume in the slab is

$$\frac{1}{v} Q_b \, \vec{d} = \sigma_b \, \hat{z} = \varepsilon_0 \vec{P}$$

[4]Note that in the dielectric, we have

$$\vec{D} = \vec{E} + \vec{P} = (1 + \chi_e)\vec{E} = \kappa \vec{E}$$

where $\vec{D} = \vec{E}_{\text{app}}$ is the applied field, and κ is the *dielectric constant.*

Chapter 8

Currents and Ohm's Law

Problem 8.1 A voltage of $100\,\mathrm{V}$ across a $10^5\,\Omega$ resistor produces what current (in mA)?

Solution to Problem 8.1

From Ohm's law

$$\Delta V = IR$$

Hence

$$I = \frac{100\,\mathrm{V}}{10^5\,\Omega} = 10^{-3}\,\mathrm{A} = 1\,\mathrm{mA}$$

Problem 8.2 A voltage of 100 volts across two 100 ohm resistors in series produces what current? What if the resistors are in parallel?

Solution to Problem 8.2

If the two resistors are in series, the total resistance is

$$R = R_1 + R_2 = 200\,\Omega \qquad ; \text{ series}$$

Hence, in this case Ohm's law gives

$$I = \frac{\Delta V}{R} = \frac{100\,\mathrm{V}}{200\,\Omega} = 0.5\,\mathrm{A}$$

If the two resistors are in parallel, the total resistance is

$$\frac{1}{R} = \frac{1}{R_1} + \frac{1}{R_2} = \frac{1}{50\,\Omega} \qquad ; \text{ parallel}$$

Hence, in this case Ohm's law gives

$$I = \frac{\Delta V}{R} = \frac{100\,\text{V}}{50\,\Omega} = 2\,\text{A}$$

Chapter 9

DC Circuits

Problem 9.1 Show that Kirchoff's second rule for the *outer loop* in Fig. 9.7 is satisfied by the solution in section 9.5.2.

Solution to Problem 9.1

The solution in section 9.5.2 is obtained by applying Kirchoff's second rule to the two *inner* loops in Fig. 9.7. If Kirchoff's second rule is applied to the *outer* loop, one has

$$\mathcal{E}_1 - \mathcal{E}_2 = R_1 i_1 + R_3 i_2$$

This is indeed satisfied by the obtained solution

$$12\,\mathrm{V} - 8\,\mathrm{V} = 4\Omega\left(\frac{5}{4}\,\mathrm{A}\right) + 2\Omega\left(-\frac{1}{2}\,\mathrm{A}\right)$$

Problem 9.2 Consider the multi-loop circuit in Fig. 9.7 and Eqs. (9.18). Show the net power exerted by the batteries through Eq. (9.11) is equal to the power dissipated in the resistors in Eq. (9.10).

Solution to Problem 9.2

Consider the multiloop circuit in Fig. 9.7. The power exerted *by* the batteries through Eq. (9.11) is[1]

$$P = i_1\mathcal{E}_1 - i_2\mathcal{E}_2 \qquad ;\ \text{batteries}$$

With the use of the solution in Eqs. (9.21), one has

$$P = 12\,\mathrm{V}\left(\frac{5}{4}\mathrm{A}\right) - 8\,\mathrm{V}\left(-\frac{1}{2}\mathrm{A}\right) = 19\,\mathrm{W}$$

[1]The minus sign is because the current i_2, as defined, is pumped *through* the battery.

Here the unit of power is

$$1\,\text{watt} \equiv 1\,\text{W} = 1\,\text{VA} = 1\,\Omega\text{A}^2$$

The total power dissipated *in* the resistors is from Eq. (9.10)

$$P = i_1^2 R_1 + (i_1 - i_2)^2 R_2 + i_2^2 R_3 \qquad ;\ \text{resistors}$$

This gives

$$P = 4\Omega\left(\frac{5}{4}\text{A}\right)^2 + 4\Omega\left(\frac{7}{4}\text{A}\right)^2 + 2\Omega\left(\frac{1}{2}\text{A}\right)^2$$
$$= \left[\frac{25}{4} + \frac{49}{4} + \frac{2}{4}\right]\text{W} = 19\,\text{W}$$

Although perhaps not evident at the outset, this agrees with the power supplied by the batteries, as guaranteed by *energy conservation.*

Problem 9.3 (a) In Fig. 9.7, what must $\mathcal{E}_2$ be to make $i_2 = -1\,\text{A}$?
(b) What is the corresponding i_1?

Solution to Problem 9.3

(a) Consider the multi-loop circuit in Fig. 9.7. Suppose that instead of $\mathcal{E}_2 = 8\,\text{V}$, one makes

$$\mathcal{E}_2 = 10\,\text{V}$$

Equations (9.20), derived from Kirchoff's rules, then read

$$8\Omega\, i_1 - 4\Omega\, i_2 = 12V$$
$$-8\Omega\, i_1 + 12\Omega\, i_2 = -20V$$

The solution is now

$$i_2 = -1\,\text{A}$$

(b) The corresponding result for i_1 is

$$i_1 = 1\,\text{A}$$

Chapter 10

Review of Electricity

Problem 10.1 Two equal charges q sit at the points $(\pm d/2, 0)$ in the (x, y)-plane.

(a) Show the field along the y-axis is

$$\vec{E} = \frac{q}{2\pi\varepsilon_0} \frac{y}{(y^2 + d^2/4)^{3/2}} \,\hat{y}$$

(b) Show the field for $x > d/2$ along the positive x-axis is

$$\vec{E} = \frac{q}{2\pi\varepsilon_0} \frac{x^2 + d^2/4}{(x^2 - d^2/4)^2} \,\hat{x}$$

Solution to Problem 10.1

(a) Along the $\hat{y}$-axis, the components of the field in the $\hat{x}$-direction cancel, and the components in the $\hat{y}$-direction add. Thus, along the positive $\hat{y}$-axis

$$\vec{E} = \frac{q}{4\pi\varepsilon_0} \frac{2\cos\chi}{(y^2 + d^2/4)} \,\hat{y} \qquad ; \; \cos\chi = \frac{y}{(y^2 + d^2/4)^{1/2}}$$

Hence

$$\vec{E} = \frac{q}{2\pi\varepsilon_0} \frac{y}{(y^2 + d^2/4)^{3/2}} \,\hat{y}$$

(b) For $x > d/2$ along the positive $\hat{x}$-axis, the electric fields simply add

$$\vec{E} = \frac{q}{4\pi\varepsilon_0} \left[\frac{1}{(x - d/2)^2} + \frac{1}{(x + d/2)^2} \right] \hat{x}$$

Thus

$$\vec{E} = \frac{q}{2\pi\varepsilon_0} \frac{x^2 + d^2/4}{(x^2 - d^2/4)^2} \hat{x}$$

Problem 10.2 A point charge $+q$ sits at the center of a metallic conducting sphere of inner radius a and outer radius b.

(a) What is the electric field for $0 < r < a$?

(b) What is the electric field for $a < r < b$?

(c) What is the electric field for $r > b$?

Solution to Problem 10.2

From the symmetry of the problem, the electric field is here of the form

$$\vec{E} = E_r \hat{r}$$

(a) A gaussian sphere surrounding the point charge, with $r < a$ so that it is *inside* of the metallic conducting sphere, gives the field of the point charge

$$\vec{E} = \frac{q}{4\pi\varepsilon_0} \frac{\hat{r}}{r^2} \qquad ; r < a$$

(b) There is no field inside the conductor. This implies a total charge $-q$ must be induced on the *inside* of the metallic conducting sphere, so that a gaussian sphere at a radius r satisfying $a < r < b$ produces no result

$$\vec{E} = 0 \qquad ; a < r < b$$

(c) By charge conservation, a corresponding cancelling charge $+q$ must be induced on the *outside* of the neutral metallic conducting sphere. A gaussian sphere with $r > b$ then reproduces the result in part (a)

$$\vec{E} = \frac{q}{4\pi\varepsilon_0} \frac{\hat{r}}{r^2} \qquad ; r > b$$

Problem 10.3 A sphere of radius R and total charge Q has the charge uniformly distributed throughout.

(a) Use Gauss' law to show that the electric field inside the sphere is given by

$$\vec{E} = \frac{Q}{4\pi\varepsilon_0} \frac{r}{R^3} \hat{r} \qquad ; r < R$$

(b) Show that outside the sphere the electrostatic potential is that of a point charge, so that $V(R) = Q/4\pi\varepsilon_0 R$;

(c) Integrate the field from R to r to show the potential *inside* the sphere is[1]

$$\begin{aligned} V(r) &= V(R) - \int_R^r \vec{E} \cdot d\vec{r} \\ &= \frac{Q}{8\pi\varepsilon_0 R}\left[3 - \left(\frac{r}{R}\right)^2\right] \qquad ; \ r < R \end{aligned}$$

Solution to Problem 10.3

(a) Here the electric field is radial, with $\vec{E} = E_r\,\hat{r}$, and the constant charge density in the uniformly-charged sphere is

$$\rho = \frac{Q}{4\pi R^3/3}$$

Gauss' theorem applied to a gaussian sphere of radius $r < R$ *inside* the sphere then gives

$$4\pi r^2 E_r = \frac{1}{\varepsilon_0}\frac{4\pi r^3}{3}\rho = \frac{Q}{\varepsilon_0}\frac{r^3}{R^3}$$

Hence, inside the sphere,

$$\vec{E} = \frac{Q}{4\pi\varepsilon_0}\frac{r}{R^3}\,\hat{r} \qquad ; \ r < R$$

(b) Either from the above, or from Gauss' law outside the sphere, the field and potential at the surface, with $V(\infty) = 0$ as ground, are the point expressions

$$\begin{aligned} \vec{E} &= \frac{Q}{4\pi\varepsilon_0}\frac{1}{R^2}\hat{r} \qquad ; \ r = R \\ V(r) &= \frac{Q}{4\pi\varepsilon_0}\frac{1}{R} \end{aligned}$$

(c) Now do the integral

$$-\int_R^r \vec{E} \cdot d\vec{r} = -\int_R^r E_r\, dr = \frac{Q}{8\pi\varepsilon_0 R^3}\left(R^2 - r^2\right)$$

---

[1]Note the harmonic oscillator potential energy seen by an oppositely charged particle moving inside the sphere.

Hence

$$V(r) = V(R) - \int_R^r \vec{E} \cdot d\vec{r}$$

$$= \frac{Q}{8\pi\varepsilon_0 R}\left[3 - \left(\frac{r}{R}\right)^2\right] \qquad ; r < R$$

PART 2
Magnetism

Chapter 11

Vectors

Problem 11.1 (a) There is another way of writing the vector product of two vectors. Introduce the completely antisymmetric Levi-Civita tensor

$$\begin{aligned} \varepsilon_{ijk} &= +1 \qquad ; \ (i,j,k) \text{ an even permutation of } (1,2,3) \\ &= -1 \qquad ; \ (i,j,k) \text{ an odd permutation of } (1,2,3) \\ &= 0 \qquad ; \ \text{otherwise} \end{aligned}$$

together with the summation convention that repeated Latin indices are summed from 1 to 3. Label the (x, y, z)-axes with $(1, 2, 3)$. Show that

$$(\vec{a} \times \vec{b})_i = \varepsilon_{ijk} a_j b_k$$

(b) Show

$$\varepsilon_{ijk}\varepsilon_{ilm} = \delta_{jl}\delta_{km} - \delta_{jm}\delta_{kl}$$

where δ_{ij} is the Kronecker delta

$$\begin{aligned} \delta_{ij} &= 1 \qquad ; \ \text{if } i = j \\ &= 0 \qquad ; \ \text{if } i \neq j \end{aligned}$$

(c) Hence, establish the vector identity

$$(\vec{a} \times \vec{b}) \cdot (\vec{c} \times \vec{d}) = (\vec{a} \cdot \vec{c})(\vec{b} \cdot \vec{d}) - (\vec{a} \cdot \vec{d})(\vec{b} \cdot \vec{c})$$

(d) Derive the vector identity

$$\vec{a} \times (\vec{b} \times \vec{c}) = (\vec{a} \cdot \vec{c})\,\vec{b} - (\vec{a} \cdot \vec{b})\,\vec{c}$$

Solution to Problem 11.1

(a) There are 6 non-zero elements to the Levi-Civita tensor

$$\varepsilon_{123} = \varepsilon_{312} = \varepsilon_{231} = +1$$
$$\varepsilon_{213} = \varepsilon_{321} = \varepsilon_{132} = -1$$

With the summation convention, one has

$$\vec{a} \cdot \vec{b} = a_i b_i = a_1 b_1 + a_2 b_2 + a_3 b_3$$

The first component of the indicated cross-product is then given by

$$\begin{aligned}(\vec{a} \times \vec{b})_1 &= \varepsilon_{1jk} a_j b_k = \varepsilon_{123} a_2 b_3 + \varepsilon_{132} a_3 b_2 \\ &= a_2 b_3 - a_3 b_2\end{aligned}$$

and similarly for the others. In cartesian notation, this reproduces the expression in Eq. (11.10)

$$\vec{a} \times \vec{b} \equiv (a_y b_z - a_z b_y)\,\hat{x} + (a_z b_x - a_x b_z)\,\hat{y} + (a_x b_y - a_y b_x)\,\hat{z} \quad ;\ \text{vector product}$$

(b) Write out

$$\varepsilon_{ijk}\varepsilon_{ilm} = \varepsilon_{1jk}\varepsilon_{1lm} + \varepsilon_{2jk}\varepsilon_{2lm} + \varepsilon_{3jk}\varepsilon_{3lm}$$

Take any pair of values for (j, k), say $(2, 3)$; they must be different. In this case, only the first term on the r.h.s. $\varepsilon_{1jk}\varepsilon_{1lm}$ contributes. The second pair of indices (l, m) is then also, by necessity, confined to $(2, 3)$; they must also be different. Suppose $j = l$ and $k = m$, the contribution of this term is

$$\varepsilon_{123}\,\varepsilon_{123} = \varepsilon_{132}\,\varepsilon_{132} = +1$$

Suppose $j = m$ and $k = l$, the contribution of this term is

$$\varepsilon_{123}\,\varepsilon_{132} = \varepsilon_{132}\,\varepsilon_{123} = -1$$

Now within this $(2, 3)$-subspace, there are a total of $2^4 = 16$ terms for $\varepsilon_{1jk}\varepsilon_{1lm}$, and only the above 4 are non-zero. The following expression reproduces these results

$$\begin{aligned}\delta_{jl}\delta_{km} - \delta_{jm}\delta_{kl} &= 0 && ;\ j = k,\ \underline{\text{or}},\ l = m \\ &= +1 && ;\ j = l,\ k = m \\ &= -1 && ;\ j = m,\ k = l\end{aligned}$$

The other two subspaces, $(1,2)$ and $(3,1)$, are handled by the additional two terms on the r.h.s. of the first expression above in part (b).[1] Hence

$$\varepsilon_{ijk}\varepsilon_{ilm} = \delta_{jl}\delta_{km} - \delta_{jm}\delta_{kl}$$

(c) Use the above to write

$$\begin{aligned}(\vec{a}\times\vec{b})\cdot(\vec{c}\times\vec{d}) &= \varepsilon_{ijk}\,\varepsilon_{ilm}\,a_j b_k c_l d_m \\ &= [\delta_{jl}\delta_{km} - \delta_{jm}\delta_{kl}]\;a_j b_k c_l d_m \\ &= (\vec{a}\cdot\vec{c})(\vec{b}\cdot\vec{d}) - (\vec{a}\cdot\vec{d})(\vec{b}\cdot\vec{c})\end{aligned}$$

(d) In a similar fashion, write

$$\begin{aligned}\left[\vec{a}\times(\vec{b}\times\vec{c})\right]_j &= \varepsilon_{jki}\,a_k\,(\varepsilon_{ilm}\,b_l c_m) = \varepsilon_{ijk}\,\varepsilon_{ilm}\,a_k b_l c_m \\ &= [\delta_{jl}\delta_{km} - \delta_{jm}\delta_{kl}]\,a_k b_l c_m \\ &= (\vec{a}\cdot\vec{c})\,b_j - (\vec{a}\cdot\vec{b})\,c_j\end{aligned}$$

In vector form this reads

$$\vec{a}\times(\vec{b}\times\vec{c}) = (\vec{a}\cdot\vec{c})\,\vec{b} - (\vec{a}\cdot\vec{b})\,\vec{c}$$

Problem 11.2 Show that the vector triple product is invariant under cyclic permutations

$$\vec{a}\cdot(\vec{b}\times\vec{c}) = \vec{c}\cdot(\vec{a}\times\vec{b}) = \vec{b}\cdot(\vec{c}\times\vec{a})$$

Solution to Problem 11.2

It is readily established that the Levi-Civita tensor is invariant under a cyclic permutation of its indices

$$\varepsilon_{ijk} = \varepsilon_{kij} = \varepsilon_{jki}$$

Hence

$$\begin{aligned}\vec{a}\cdot(\vec{b}\times\vec{c}) &= \varepsilon_{ijk}a_i b_j c_k \\ &= \varepsilon_{kij}c_k a_i b_j = \varepsilon_{jki}b_j c_k a_i\end{aligned}$$

It follows that

$$\vec{a}\cdot(\vec{b}\times\vec{c}) = \vec{c}\cdot(\vec{a}\times\vec{b}) = \vec{b}\cdot(\vec{c}\times\vec{a})$$

[1] This encompasses all the non-zero terms in $\delta_{jl}\delta_{km} - \delta_{jm}\delta_{kl}$. If you are unconvinced, just write out the $3^4 = 81$ possibilities for (j,k,l,m) in $\varepsilon_{ijk}\varepsilon_{ilm} = \delta_{jl}\delta_{km} - \delta_{jm}\delta_{kl}$.

Problem 11.3 Show that a vector $\vec{v} = v_x\,\hat{x} + v_y\,\hat{y} + v_z\,\hat{z}$ can also be written as a length times a unit vector denoting the direction $\vec{v} = v\hat{n}$. What are $(v, \hat{n})$?

Solution to Problem 11.3

The square of the length of the vector is[2]

$$v^2 = \vec{v}\cdot\vec{v} = v_x^2 + v_y^2 + v_z^2$$

The length is therefore

$$v = (v_x^2 + v_y^2 + v_z^2)^{1/2}$$

The quantity $\hat{n} \equiv \vec{v}/v$ is then a *unit vector* denoting the direction

$$\hat{n} \equiv \frac{\vec{v}}{v} = \frac{1}{(v_x^2 + v_y^2 + v_z^2)^{1/2}}(v_x\,\hat{x} + v_y\,\hat{y} + v_z\,\hat{z})$$

The vector itself is thus written as

$$\vec{v} = v_x\,\hat{x} + v_y\,\hat{y} + v_z\,\hat{z} = v\,\hat{n}$$

Problem 11.4 Show that if $\vec{r}$ is the position vector in three-dimensions, the *divergence* gives $\vec{\nabla}\cdot\vec{r} = 3$. Show the *gradient* satisfies $\vec{\nabla}f(r) = (\hat{r}\,\partial/\partial r)f(r)$.

Solution to Problem 11.4

In three-dimensions, the vector $\vec{r}$ is given by

$$\vec{r} = x\,\hat{x} + y\,\hat{y} + z\,\hat{z}$$

The divergence of this vector is

$$\vec{\nabla}\cdot\vec{r} = \frac{\partial x}{\partial x} + \frac{\partial y}{\partial y} + \frac{\partial z}{\partial z} = 3$$

For the second part, we can just follow the argument in section 5.9.2

$$\begin{aligned}\vec{\nabla}f(r) &= \left(\hat{x}\frac{\partial}{\partial x} + \hat{y}\frac{\partial}{\partial y} + \hat{z}\frac{\partial}{\partial z}\right)f(r) \qquad ;\ r = (x^2 + y^2 + z^2)^{1/2} \\ &= \frac{\partial f(r)}{\partial r}\frac{1}{(x^2 + y^2 + z^2)^{1/2}}(x\,\hat{x} + y\,\hat{y} + z\,\hat{z}) \\ &= \frac{\partial f(r)}{\partial r}\frac{\vec{r}}{r}\end{aligned}$$

[2] Note $|\vec{v}|^2 \equiv \vec{v}\cdot\vec{v} = \vec{v}^2 = v^2$.

Thus

$$\vec{\nabla} f(r) = \left(\hat{r} \frac{\partial}{\partial r} \right) f(r)$$

Chapter 12

The Magnetic Force and Field

Problem 12.1 Suppose one sits a distance $h/2 + \varepsilon$ above the bottom Helmholtz coil in Fig. 12.11, where the coils are a distance h apart. Show that to leading order in the small displacement from the midpoint, the magnetic field along the axis is[1]

$$\vec{B} = \frac{\mu_0 i}{2\pi}(\pi R^2)\frac{2}{d^3}\left\{1 + \frac{3}{2}\left(\frac{\varepsilon}{d}\right)^2\left[\frac{5}{4}\left(\frac{h}{d}\right)^2 - 1\right]\right\}\hat{z} \qquad ; \; d^2 \equiv R^2 + \left(\frac{h}{2}\right)^2$$

Solution to Problem 12.1

The general expression for the field along the axis of a *current loop*, derived in the text from the Biot-Savart law, is given in Eq. (12.23)

$$\vec{B} = \frac{\mu_0 i}{2\pi}\frac{\pi R^2}{(z^2 + R^2)^{3/2}}\,\hat{z} \qquad ; \text{ along z-axis}$$

If one sits a distance $h/2 + \varepsilon$ above the bottom Helmholtz coil in Fig. 12.11, where the coils are a distance h apart, then by superposition

$$\vec{B} = \frac{\mu_0 i}{2\pi}(\pi R^2)\hat{z}\, I$$
$$I \equiv \frac{1}{[(h/2 + \varepsilon)^2 + R^2]^{3/2}} + \frac{1}{[(h/2 - \varepsilon)^2 + R^2]^{3/2}}$$

[1]Recall Eqs. (12.23) and (3.6).

This is re-written as

$$I = [d^2 + \varepsilon h + \varepsilon^2]^{-3/2} + [d^2 - \varepsilon h + \varepsilon^2]^{-3/2} \qquad ; \; d^2 \equiv R^2 + \left(\frac{h}{2}\right)^2$$

$$= \frac{1}{d^3}\left[\left(1 + \frac{\varepsilon h}{d^2} + \frac{\varepsilon^2}{d^2}\right)^{-3/2} + \left(1 - \frac{\varepsilon h}{d^2} + \frac{\varepsilon^2}{d^2}\right)^{-3/2}\right]$$

Now use the binomial theorem in Eq. (3.6), which holds for all n with $|x| < 1$,

$$(1+x)^n = 1 + nx + \frac{n(n-1)}{2!}x^2 + \cdots$$

It follows that

$$I = \frac{1}{d^3}\left[1 - \frac{3}{2}\left(\frac{\varepsilon h}{d^2} + \frac{\varepsilon^2}{d^2}\right) + \left(-\frac{3}{2}\right)\left(-\frac{5}{2}\right)\frac{1}{2}\left(\frac{\varepsilon h}{d^2} + \frac{\varepsilon^2}{d^2}\right)^2 + \right.$$
$$\left. 1 - \frac{3}{2}\left(-\frac{\varepsilon h}{d^2} + \frac{\varepsilon^2}{d^2}\right) + \left(-\frac{3}{2}\right)\left(-\frac{5}{2}\right)\frac{1}{2}\left(-\frac{\varepsilon h}{d^2} + \frac{\varepsilon^2}{d^2}\right)^2 + \cdots\right]$$

If terms up through $O(\varepsilon^2)$ are retained

$$I \approx \frac{1}{d^3}\left[2 - 3\left(\frac{\varepsilon}{d}\right)^2 + \frac{15}{4}\left(\frac{\varepsilon h}{d^2}\right)^2\right]$$

The field along the axis of the Helmholtz coils in this case is therefore

$$\vec{B} = \frac{\mu_0 i}{2\pi}(\pi R^2)\frac{2}{d^3}\left\{1 + \frac{3}{2}\left(\frac{\varepsilon}{d}\right)^2\left[\frac{5}{4}\left(\frac{h}{d}\right)^2 - 1\right]\right\}\hat{z} \qquad ; \; d^2 \equiv R^2 + \left(\frac{h}{2}\right)^2$$

The correction to the field is quadratic in ε, the small vertical displacement from the midpoint of the coils.[2]

[2]Even this correction vanishes if the coils are arranged so that $(h/d)^2 = 4/5$.

Chapter 13

Ampere's Law

Problem 13.1 Suppose the two ends of the infinite solonoid in Fig. 13.6 are joined to form a *torus*. Present an argument that the magnetic field is confined to the interior of the torus.

Solution to Problem 13.1

Magnetic field lines do not end.[1] As explained in footnote 5 on p. 117 of the text, what actually happens with the finite-length solonoid is that the field exits the solonoid and returns to go though it at the other end, with vanishing flux everywhere outside, if the solonoid is long enough. With a *torus*, the field lines need *never exit the solonoid*, and they can remain confined to the interior of the torus in a never-ending manner.

Problem 13.2 $\mathcal{N}$ turns of a wire carrying a current i are wrapped closely around the surface of a torus of inner radius R_1 and outer radius R_2 lying in the (x, y)-plane. The current flows in a clockwise direction when viewed in the azimuthal $\hat{\phi}$-direction.

(a) Let C be a circle of radius r in the (x, y)-plane. Use Ampere's law to show the magnetic field for $R_1 < r < R_2$ is given by

$$\vec{B} = \left(\frac{\mu_0}{2\pi}\right)\mathcal{N}i\frac{\hat{\phi}}{r}$$

(b) What is the magnetic field for $r > R_2$?

(c) If R_2 and R_1 become very large with $R_2 - R_1 \equiv d$ fixed, show that the result in (a) reproduces the correct result for an infinite, linear solonoid.

[1] In analogy to the electric field lines, the magnetic field lines are also lines of force on a magnetic charge; however, since there is no free magnetic charge, the magnetic field lines do not end, and this magnetic force can only be observed indirectly through the torque on a dipole. More generally, the unit vector $\hat{B}(\vec{r})$ is tangent to the magnetic field line at each point (compare the solution to Prob. 3.2).

Solution to Problem 13.2

(a) From symmetry, the field in the torus has the form

$$\vec{B} = B_r\,\hat{\phi}$$

Ampere's law applied to an amperian circle lying in the torus then gives

$$\oint_C \vec{B}\cdot d\vec{l} = 2\pi r B_r = \mu_0 \mathcal{N} i$$

Here $\mathcal{N}i$ is the total current flowing through the amperian loop. Hence

$$\vec{B} = \left(\frac{\mu_0}{2\pi}\right)\mathcal{N}i\frac{\hat{\phi}}{r} \qquad ; R_1 < r < R_2$$

(b) If the radius of the circle is greater than R_2, there is *no net current* flowing through the amperian loop and the field *vanishes*[2]

$$\vec{B} = 0 \qquad ; R_2 < r$$

(c) Suppose R_2 and R_1 become very large with $R_2 - R_1 \equiv d$ fixed. In this case we can write

$$\frac{\mathcal{N}}{2\pi r} \approx \frac{\mathcal{N}}{2\pi R_1} \equiv N \qquad ; \text{ turns/length}$$

where N is the number of turns per unit length in the solonoid. If we define $\hat{\phi} \equiv \hat{n}$, where $\hat{n}$ is a unit vector that points down the axis of the torus, then the magnetic field in the torus becomes

$$\begin{aligned} \vec{B} &= \mu_0 N i\,\hat{n} \qquad &&; \text{ inside torus} \\ &= 0 &&; \text{ outside torus} \end{aligned}$$

This is the correct result for an infinite, linear solonoid in Eq. (13.19).

Problem 13.3 Given two identical, parallel, current line elements $id\vec{l}$ separated by a perpendicular distance $\vec{r}$. Use the Biot-Savart law in Eq. (12.8), the element of force in Eq. (13.40), and the vector identity in Prob. 11.1(d) to show that the second current element is attracted to the first with a force

$$d\vec{F}_{12} = \frac{\mu_0}{4\pi r^2}(id\vec{l}\,)^2(-\hat{r})$$

[2]The same argument holds if $r < R_1$.

Solution to Problem 13.3

The Biot-Savart law in Eq. (12.8) states that

$$d\vec{B} = \frac{\mu_0}{4\pi}\frac{id\vec{l}\times\hat{r}}{r^2} \qquad \text{; Biot-Savart law}$$

The element of force in Eq. (13.40) is

$$d\vec{F} = i\,d\vec{l}\times\vec{B} \qquad \text{; force on current element}$$

Hence, with the aid of the vector identity in Prob. 11.1(d), the attractive force between two parallel current elements $id\vec{l}$, with $\hat{r}\cdot(id\vec{l}\,)=0$, is

$$\begin{aligned} d\vec{F}_{12} &= \frac{\mu_0}{4\pi}i\,d\vec{l}\times\left[\frac{id\vec{l}\times\hat{r}}{r^2}\right] \\ &= \frac{\mu_0}{4\pi r^2}(id\vec{l}\,)^2(-\hat{r}) \end{aligned}$$

Problem 13.4 More complicated magnetic field configurations can again be obtained by combining solvable components using the principle of *superposition.* For example, suppose there are two crossed lines of current i in the (x,y)-plane flowing along the x- and y-axes in the positive direction. Show the field along the positive z-axis is

$$\vec{B} = \frac{\mu_0 i}{2\pi z}\,(\hat{x}-\hat{y})$$

Solution to Problem 13.4

It is assumed that the two lines of current do not affect each other.

As in Prob. 4.7, consider the right-handed cartesian coordinate system in Fig. 11.1. The field in the $\hat{x}$ direction along the positive z-axis surrounding the current along the y-axis is

$$\vec{B}_1 = \frac{\mu_0 i}{2\pi z}\hat{x}$$

Similarly, the field in the $-\hat{y}$ direction along the positive z-axis surrounding the current along the x-axis is

$$\vec{B}_2 = -\frac{\mu_0 i}{2\pi z}\hat{y}$$

Hence, by *superposition*, the field along the positive z-axis arising from two crossed lines of current i in the (x,y)-plane flowing along the x- and

y-axes in the positive direction is

$$\vec{B} = \vec{B}_1 + \vec{B}_2 = \frac{\mu_0 i}{2\pi z}(\hat{x} - \hat{y})$$

Problem 13.5 A particle of mass and charge (m, q) starts down the axis of the solonoid in Fig. 13.6. Describe its orbit.

Solution to Problem 13.5

The magnetic field in the solonoid is uniform and points down the axis. There is no electric field. Hence if the particle moves down the axis, the Lorentz force *vanishes*

$$\vec{F} = q\left(\vec{E} + \vec{v} \times \vec{B}\right) = 0 \qquad ; \text{ down axis}$$

The particle's orbit is therefore a *straight line.*

Problem 13.6 Given the mass spectrometer result in Eq. (13.34), suppose one can measure the position to an accuracy $dR/R \sim 10^{-3}$. What is the corresponding accuracy in isotope mass dm/m?[3]

Solution to Problem 13.6

The expression in Eq. (13.34) for the charge/mass ratio for the mass spectrometer discussed in the text is

$$\frac{q}{m} = \frac{2(\Delta V)}{R^2 B^2} \equiv \frac{K}{R^2}$$

Differentiate this relation at fixed (quantized) q, and then divide by it, to obtain

$$\frac{dm}{m} = 2\frac{dR}{R}$$

Suppose one can measure the position to an accuracy $dR/R \sim 10^{-3}$. The corresponding accuracy in isotope mass dm/m is then

$$\frac{dm}{m} \sim 2 \times 10^{-3} \qquad ; \text{ if } \frac{dR}{R} \sim 10^{-3}$$

[3]Same q.

Chapter 14

Electromagnetic Induction

Problem 14.1 A loop containing 100 turns of wire is wound in the form of a square 0.1 m on each side. It is rotated with angular frequency $\omega/2\pi = 10\,\text{sec}^{-1}$ about a $\hat{y}$-axis in a uniform field $\vec{B} = 1\,\text{T}\,\hat{z}$ as shown in Fig. 14.5.

(a) Calculate the resulting voltage $\mathcal{V}$ in volts as a function of time;

(b) Sketch this voltage as a function of time. Label all axes carefully;

(c) Will this give rise to an alternating current or a direct current in an external resistance?

Solution to Problem 14.1

(a) The configuration for the AC generator is shown in Fig. 14.5. The magnetic flux through the coil, and induced EMF in the circuit, are given in Eqs. (14.8)–(14.9)

$$\Phi_m = NAB\cos\omega t$$
$$-\frac{d\Phi_m}{dt} = \omega NAB\sin\omega t = (\mathcal{EMF})_{\text{ind}}$$

We are given the following parameters

$$B = 1\,\text{T} \qquad ; A = 10^{-2}\,\text{m}^2 \qquad ; N = 100 \qquad ; \omega = 20\pi\,\text{s}^{-1}$$

Hence

$$(\mathcal{EMF})_{\text{ind}} = 100 \times 10^{-2}\,\text{m}^2 \times 1\,\text{T} \times 20\pi\,\text{s}^{-1} \times \sin(20\pi t)$$

where t is measured in sec. Use the unit conversion

$$1\,\text{V} = \frac{1\,\text{J}}{1\,\text{C}} = \frac{1\,\text{Nm}}{1\,\text{C}} = \frac{1\,\text{m}}{1\,\text{C}}\,\frac{1\,\text{CmT}}{1\,\text{s}} = \frac{1\,\text{Tm}^2}{1\,\text{s}}$$

It follows that the resulting voltage is

$$\mathcal{V} \equiv (\mathcal{E}MF)_{\rm ind} = 20\pi \sin(20\pi t) \text{ V}$$

(b) The resulting output voltage is just that of Fig. 14.6, where the labels on the axes are

$$\omega NAB = 20\pi \text{ V} \qquad ; \frac{2\pi}{\omega} = 0.10 \text{ sec}$$

(c) The current in an external resistance of R ohms in amperes A is

$$i = \frac{\mathcal{V}}{\rm R} = \frac{20\pi}{\rm R} \sin(20\pi t) \text{ A}$$

This is an *alternating current.*

Problem 14.2 In the caption to Fig. 14.11 on the DC motor, it is stated that the torque on the rectangular armature is $\vec{\mu} \times \vec{B}$, where $\vec{\mu} = i(N\mathcal{A})\hat{n}$. Here N is the number of turns, and $\mathcal{A}$ is the area of the rectangle. Verify this statement.

Solution to Problem 14.2

Suppose the loop in Fig. 14.9 is a rectangle with sides (l_1, l_2). The torque in Eq. (14.10) then becomes

$$|\vec{\tau}| = 2(il_1B)\left(\frac{l_2}{2}\sin\phi\right) = i(l_1l_2)B\sin\phi$$

In vector form, the torque is

$$\vec{\tau} = \vec{\mu} \times \vec{B} \qquad ; \text{ torque on current loop}$$
$$\vec{\mu} \equiv i(l_1l_2)\,\hat{n} = i(\mathcal{A})\,\hat{n}$$

where $\mathcal{A} = l_1l_2$ is the area of the loop. If there are N loops, the torques add. Hence, the torque on the rectangular armature in Fig. 14.11 is $\vec{\mu} \times \vec{B}$, where $\vec{\mu} = i(N\mathcal{A})\hat{n}$.

Problem 14.3 In analogy to Fig. 14.11, design an AC motor.

Solution to Problem 14.3

The torque on the current loop in the DC motor is given by (see the previous problem)

$$\vec{\tau} = \vec{\mu} \times \vec{B} \qquad ; \text{ torque on current loop}$$
$$\vec{\mu} \equiv il^2\,\hat{n} = i(\mathcal{A})\,\hat{n}$$

The direction of the loop current and $\hat{n}$ are related by the right-hand rule: if your fingers point in the direction of the current, your thumb points in the direction of $\hat{n}$.

In the discussion of Fig. 14.11, it states that, "The loop feels a torque, which turns it. If a split-ring commutator is used to reverse the input leads just as $\hat{n} \times \vec{B}$ changes sign, then the torque on the loop will always be in the same direction and the device will indeed act as a motor which rotates about its axis."

There is another way to reverse the current just as $\hat{n} \times \vec{B}$ changes sign, and that is, instead of a DC battery and a split-ring commutator, to have an alternating current EMF source with brushes (see Fig. 14.5) operating with an angular frequency ω matched to the angular frequency of the rotor. This produces an AC motor.[1]

Problem 14.4 A solonoid of radius 1 cm is wrapped with wire at 10 turns per cm. What is its self-inductance per unit length in H/m?

Solution to Problem 14.4

The self-inductance per unit length of a single solonoid is given by Eq. (14.19) as

$$\frac{1}{l}L = \mu_0 n^2(\pi r^2) \qquad \text{; single solonoid}$$

With the aid of Prob. 16.2, one then has

$$\begin{aligned}\frac{1}{l}L &= \left(4\pi \times 10^{-7}\,\text{Ns}^2/\text{C}^2\right)\left(10^3/\text{m}\right)^2\left[\pi(10^{-2}\,\text{m})^2\right] \\ &= 3.95 \times 10^{-4}\,\text{H/m}\end{aligned}$$

[1]A small DC motor can be used to bring the rotor of the AC motor up to speed.

Chapter 15

Magnetic Materials

Problem 15.1 With paramagnetic materials, one can introduce the local *magnetization* arising from the oriented dipoles in sections 15.2–15.3

$$\vec{M} \equiv \mu_0 \eta_{\text{surface}}\, \hat{n} \qquad ; \text{ magnetization}$$

(a) In general, the magnetization will be proportional to the applied field, and the constant of proportionality is known as the *susceptibility*

$$\vec{M} = \chi_m\, \vec{H} \qquad ; \text{ magnetic susceptibility}$$

Show that the magnetic field in the material is then given by

$$\begin{aligned} \vec{B} &= \vec{H} + \vec{M} = \kappa_m \vec{H} \\ \kappa_m &= 1 + \chi_m \end{aligned}$$

(b) Suppose the sample in Fig. 15.5 is of finite length. The differential form of Gauss' law for the magnetic field is derived in Eqs. (18.16)

$$\vec{\nabla} \cdot \vec{B} = 0$$

Show from part (a), that this implies

$$\vec{\nabla} \cdot \vec{H} = -\vec{\nabla} \cdot \vec{M} \equiv \mu_0\, \rho_m$$

where the r.h.s. defines ρ_m as an *effective* magnetic charge density. In the absence of free currents the static applied field satisfies [see Eqs. (18.20)]

$$\vec{\nabla} \times \vec{H} = 0$$

The corresponding equations in electrostatics are

$$\vec{\nabla} \cdot \vec{E} = \frac{\rho_e}{\varepsilon_0}$$
$$\vec{\nabla} \times \vec{E} = 0$$

where ρ_e is the electric charge density. All of electrostatics follows from these equations.

Solution to Problem 15.1

(a) It is clear from the discussion in section 15.2 that in the material, the magnetic field gets a contribution from both the applied field and the oriented dipoles, the latter being described by a surface current per unit length η_{surface}. Define the magnetization $\vec{M}$ *in* the material by

$$\vec{M} \equiv \mu_0 \eta_{\text{surface}}\, \hat{n} \qquad ; \text{ magnetization}$$

Let $\vec{H}$ be the applied field

$$\vec{H} \equiv \vec{B}_{\text{app}} = \mu_0 \eta_{\text{free}}\, \hat{n} \qquad ; \text{ defines } \vec{H}$$

Equations (15.10) and (15.15) then state that in the material

$$\vec{B} = \vec{H} + \vec{M} \qquad ; \text{ in material}$$

In general, the magnetization will be proportional to the applied field

$$\vec{M} = \chi_m \vec{H} \qquad ; \text{ magnetic susceptibility}$$

It follows that in the material [see Eq. (15.11)].

$$\vec{B} = \vec{H} + \vec{M} = \kappa_m \vec{H}$$
$$\kappa_m = 1 + \chi_m$$

(b) Now suppose the sample in Fig. 15.5 is of *finite length*. The differential form of Gauss' law for the magnetic field is derived in Eqs. (18.16)

$$\vec{\nabla} \cdot \vec{B} = 0$$

There is no free magnetic charge, and magnetic field lines of $\vec{B}$ do not end. Take the divergence of the last expression in part (a) to obtain

$$\vec{\nabla} \cdot \vec{H} = -\vec{\nabla} \cdot \vec{M} \equiv \mu_0\, \rho_m$$

Here the last relation defines $\mu_0\, \rho_m \equiv -\vec{\nabla} \cdot \vec{M}$ as an *effective* magnetic charge density.

The source of the static applied field in part (a) is the free current, and if there are no free currents, then [see Eqs. (18.22) and the solution to Prob. 23.4(a)]

$$\vec{\nabla} \times \vec{H} = 0 \qquad ; \vec{j}_{\text{free}} = 0$$

The corresponding equations in electrostatics are

$$\vec{\nabla} \cdot \vec{E} = \frac{\rho_e}{\varepsilon_0}$$

$$\vec{\nabla} \times \vec{E} = 0$$

where ρ_e is the electric charge density. All of electrostatics follows from these equations. With a static situation, and no free current, there is a clear analogy ($\vec{E} \leftrightharpoons \vec{H}$) and ($\rho_e/\varepsilon_0 \leftrightharpoons \mu_0\rho_m$).

Problem 15.2[1] Now suppose the sample in Prob. 15.1 is a *permanent magnet*, where the magnetization $\vec{M}$ in the sample remains finite as the applied field goes to zero, so that in the sample in Fig. 15.5, $\vec{H} = 0$ and $\vec{B} = \vec{M}$.

(a) Discuss how one can use the analogy to *electrostatics* to calculate the magnetic field outside a finite permanent magnet in *magnetostatics*.[2]

(b) Sketch the field outside a long, thin bar magnet; also, sketch the field of a *c*-shaped magnet.

Solution to Problem 15.2

(a) With a finite-length permanent magnet, there are equal and opposite effective magnetic charges at both ends, where the divergence of the magnetization is non-zero (see Prob. 15.3). In direct analogy to electrostatics, one can focus on the field $\vec{H}$ and introduce an *effective* magnetostatic potential $V_m(\vec{x})$ that receives a contribution from $\rho_m(\vec{x})$, exactly as in Prob. 5.4

$$V_m(\vec{x}) = \frac{\mu_0}{4\pi} \int \frac{\rho_m(\vec{x}\,')d^3x'}{|\vec{x} - \vec{x}\,'|}$$

The field $\vec{H}$ is then obtained as

$$\vec{H}(\vec{x}) = -\vec{\nabla}\, V_m(\vec{x})$$

[1] This problem has been re-worded slightly for clarity.

[2] Recall Prob. 5.4. Note that in magnetostatics, $\vec{M}$ is only non-zero *inside* the magnets, and the magnetic field outside can be obtained from $\vec{H}$ through $\vec{B} = \kappa_m \vec{H}$, with $\kappa_m = 1$. (It helps to remember that the normal component of $\vec{B}$ is continuous across an interface.) Notice also that there may be a *demagnetizing* field *inside* the finite-length permanent magnet due to the effective magnetic charges at the ends.

The magnetic field $\vec{B}$, which determines the magnetic force on a charged particle, is then determined from $\vec{H}$ outside the magnet, where $\kappa_m = 1$, through[3]

$$\vec{B} = \kappa_m \vec{H} = \vec{H} \qquad ; \text{ outside magnet}$$

One can now use all the tricks of the trade of *electrostatics* to discuss the *magnetostatics* of permanent magnets.

(b) See the *Aside* below.

Problem 15.3 Consider $\int \rho_m dv$ over a gaussian pillbox on the transverse face of the permanent magnet in Prob. 15.2. Use Gauss' theorem to show that on the face

$$\vec{M} = \mu_0 \sigma_m \hat{n}$$

where σ_m is an *effective* surface magnetic charge density.

Solution to Problem 15.3

Consider the permanent magnet of Prob. 15.2, and start from the relation in Prob. 15.1(b)

$$\mu_0 \rho_m = -\vec{\nabla} \cdot \vec{M}$$

The magnetization inside this magnet is constant and of the form

$$\vec{M} = M \hat{n}$$

Construct a gaussian pillbox on the transverse face of the magnet as shown in Fig. 15.1 below. Integrate the above relation over the volume of this pillbox to find the total effective magnetic charge contained in the pillbox

$$\mu_0 q_m = \mu_0 \int_V \rho_m \, dv = \mu_0 \int_V (-\vec{\nabla} \cdot \vec{M}) dv$$

Now use Gauss' theorem on the integral

$$\int_V (-\vec{\nabla} \cdot \vec{M}) dv = -\int_S \vec{M} \cdot d\vec{S}$$

[3]Although the applied field $\vec{H}$ vanishes inside the infinite sample in Fig. 5.5 when there is no free surface current, there will be a demagnetizing field inside the *finite-length* sample due to the fact that $\vec{\nabla} \cdot \vec{M}$ is non-zero at the ends; inside the finite-length permanent magnet, one again has $\vec{B} = \vec{H} + \vec{M}$ (it is assumed here that $\vec{M}$ is unaffected).

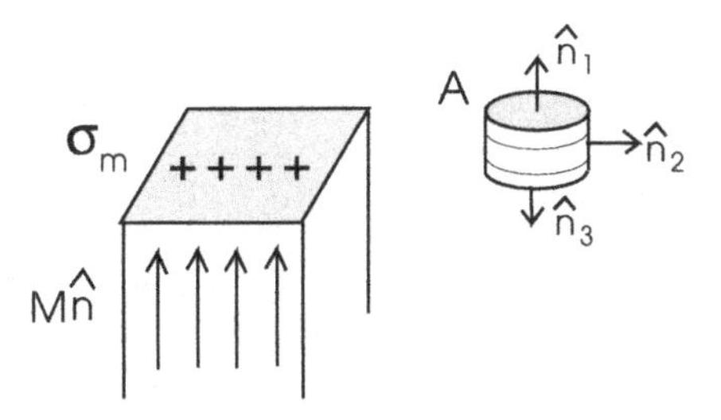

Fig. 15.1 Gaussian pillbox surrounding an element of area A on the surface of a permanent magnet where $\vec{M} = M\,\hat{n}$ in the magnet. σ_m is an *effective* magnetic surface charge density.

The surface integral only receives a contribution from the surface *inside* the magnet, and the normal to this surface points in the direction $-\hat{n}$. Hence

$$\int_S \vec{M} \cdot d\vec{S} = -M\,A$$

The physical quantity is the (positive) *effective* magnetic surface charge density $\sigma_m \equiv q_m/A$, hence

$$\mu_0\,\sigma_m \equiv \mu_0\,\frac{q_m}{A} = M$$

A combination of the above gives

$$\vec{M} = \mu_0 \sigma_m \hat{n}$$

Consider the face at the other end of the magnet, where the magnetization $\vec{M}$ points *into* the magnet. Since the normal to the contributing surface inside the magnet now points in the direction $+\hat{n}$, the gaussian pillbox gives the *effective* magnetic surface charge density on this face as $-\sigma_m$; the permanent magnet is a *magnetic dipole.*

****

(*Aside*) We give two examples of the magnetic field from permanent magnets obtained through the analysis in Probs. 15.1–15.3 [compare Prob. 15.2(b)]:

(1) Consider a long, thin bar magnet where the sources are well localized at the ends. Let $\vec{r}$ be the position of the field point relative to the midpoint of the magnet, and let the ends of the magnet be at the positions $\pm\vec{d}/2$ along the z-axis. The field outside the magnet is then precisely the dipole

field of Prob. 3.1

$$\vec{B}(\vec{r}) = \vec{H}(\vec{r}) = \frac{\mu_0\, q_m}{4\pi}\left[\frac{(\vec{r}-\vec{d}/2)}{|(\vec{r}-\vec{d}/2)|^3} - \frac{(\vec{r}+\vec{d}/2)}{|(\vec{r}+\vec{d}/2)|^3}\right] \qquad ;\ \text{dipole field}$$

(2) Consider a c-shaped permanent magnet with two large, flat, parallel pole faces. The field configuration between the poles is just that of Fig. 3.8(b), and the field is that of Eq. (3.17) (see also Prob. 4.3)

$$\vec{B} = \vec{H} = \mu_0\, \sigma_m\, \hat{z} \qquad ;\ \text{between two sheets with opposite charge}$$

From Prob. 15.3, this can be re-written as[4]

$$\vec{B} = \vec{M}$$

The field between the poles is exactly the same as that in the material. It can become very large.

****

Problem 15.4 (a) Explain the attraction and repulsion of bar magnets;
(b) Why do permanent magnets stick to your refrigerator door?

Solution to Problem 15.4

(a) It was shown in Probs. 15.1–15.3 that there is a direct analogy between the electric field $\vec{E}$ in electrostatics and the magnetic field $\vec{H}$ in magnetostatics, where the analog of the electric charge density ρ_e/ε_0 is the effective magnetic charge density $\mu_0\rho_m = -\vec{\nabla}\cdot\vec{M}$. In direct analogy, like magnetic charges repel and unlike charges attract. In free space, the magnetic field is $\vec{B} = \vec{H}$, and in a magnetic material $\vec{B} = \vec{H} + \vec{M}$.

A permanent bar magnet acts as a dipole (see the above *Aside*). When the dipoles are aligned, the like charges are close together, and they repel. When they are anti-aligned, the unlike charge are close together, and they attract.[5]

(b) Again, in direct analogy, when a magnetic charge is placed close to a magnetic material, there will be an *induced* magnetic charge in the material (that is, an induced localized *magnetization*). A magnetic dipole will induce an anti-aligned dipole in a magnetic material, and these anti-aligned dipoles attract.

[4]Here $\hat{z} = \hat{n}$, and $\vec{M}$ is the *magnetization*. With this configuration, $\vec{H}$ is only non-zero between the pole faces.

[5]When the magnets are anti-aligned, there is a convenient return path for the non-ending field lines of $\vec{B}$; when aligned, the return paths "fight each other".

Problem 15.5 For clarity in this text, we have simply defined the fields $(\vec{D}, \vec{H})$ to be the *applied fields.* In the literature, in SI units, you will find these auxiliary fields defined with the constants (ε_0, μ_0) taken out so that

$$\vec{\mathcal{D}} = \varepsilon_0 \vec{E} + \vec{\mathcal{P}}$$
$$\vec{B} = \mu_0 \left(\vec{\mathcal{H}} + \vec{\mathcal{M}}\right)$$

(a) How are the auxiliary fields $(\vec{\mathcal{D}}, \vec{\mathcal{H}})$ related to the fields $(\vec{D}, \vec{H})$ used in this text?

(b) How are the polarization and magnetization $(\vec{\mathcal{P}}, \vec{\mathcal{M}})$ related to the $(\vec{P}, \vec{M})$ introduced in Probs. 7.4 and 15.1-15.3?

(c) Show that $(\vec{\mathcal{D}}, \vec{\mathcal{P}})$ are related to $(\sigma_{\text{free}}, \sigma_{\text{bound}})$ in Fig. 7.6 and sections 7.4–7.5 by

$$\vec{\mathcal{D}} = \sigma_{\text{free}}\,\hat{z} \qquad ; \vec{\mathcal{P}} = \sigma_{\text{bound}}\,\hat{z}$$

(d) Show that $(\vec{\mathcal{H}}, \vec{\mathcal{M}})$ are related to $(\eta_{\text{free}}, \eta_{\text{surface}})$ in Fig. 15.5 and sections 15.2–15.3 by

$$\vec{\mathcal{H}} = \eta_{\text{free}}\,\hat{n} \qquad ; \vec{\mathcal{M}} = \eta_{\text{surface}}\,\hat{n}$$

Solution to Problem 15.5

(a) In the text, we simply define the fields $(\vec{D}, \vec{H})$ to be the *applied fields* so that

$$\vec{D} = \vec{E} + \vec{P}$$
$$\vec{B} = \vec{H} + \vec{M}$$

It follows from the above that then

$$\vec{\mathcal{D}} = \varepsilon_0\,\vec{D}$$
$$\mu_0\,\vec{\mathcal{H}} = \vec{H}$$

(b) In exactly the same manner, one has

$$\vec{\mathcal{P}} = \varepsilon_0\,\vec{P}$$
$$\mu_0\,\vec{\mathcal{M}} = \vec{M}$$

(c) It is shown in the text in Eq. (7.16) that

$$\vec{E} = \vec{D} - \vec{P} = \frac{1}{\varepsilon_0}(\sigma_{\text{free}} - \sigma_b)\,\hat{z}$$

Hence

$$\vec{\mathcal{D}} = \sigma_{\text{free}}\,\hat{z} \qquad ; \; \vec{\mathcal{P}} = \sigma_{\text{bound}}\,\hat{z}$$

(d) In a similar fashion, it is shown in the text in Eq. (15.10) that

$$\vec{B} = \vec{H} + \vec{M} = \mu_0\,(\eta_{\text{free}} + \eta_{\text{surface}})\,\hat{n}$$

Thus

$$\vec{\mathcal{H}} = \eta_{\text{free}}\,\hat{n} \qquad ; \; \vec{\mathcal{M}} = \eta_{\text{surface}}\,\hat{n}$$

Chapter 16

Time-Dependent Circuits

Problem 16.1 For the LC oscillator in section 16.3, the sum of the energies in the capacitor and inductor is

$$E = \frac{1}{2C}q^2 + \frac{1}{2}Li^2 \qquad ; \text{ LC oscillator}$$

Take the time derivative and show

$$\frac{dE}{dt} = 0$$

Hence, conclude that the energy is a constant of the motion. Discuss.[1]

Solution to Problem 16.1

The configuration for the LC oscillator is shown in Fig. 16.3. Here the current is related to the rate of decrease of the charge on the plates by Eq. (16.2)

$$i = -\frac{dq}{dt}$$

Kirchoff's second rule for this system gives Eq. (16.10)

$$\frac{q}{C} - L\frac{di}{dt} = 0$$

The total energy in the oscillator is given in the statement of the problem. Take the time derivative of this energy, and use the above

$$\begin{aligned}\frac{dE}{dt} &= \frac{q}{C}\frac{dq}{dt} + Li\frac{di}{dt} \\ &= i\left[L\frac{di}{dt} - \frac{q}{C}\right] = 0\end{aligned}$$

[1]Recall the footnote on p. 148.

Hence, although the energy oscillates back and forth between the condenser and inductor, the total energy in the oscillator does not change with time, and *energy is conserved.*

The argument can be turned around, and energy conservation used here to *derive* the above statement of Kirchoff's second rule

$$\frac{dE}{dt} = 0 \qquad \Longrightarrow \qquad \frac{q}{C} - L\frac{di}{dt} = 0$$

This is interesting, since, although justified in the time-independent case, our treatment of time-dependent circuits was based on a rather *ad hoc* application of this rule:

> *We will just make use of Kirchoff's second rule in Eq. (10.45), where the net EMF in the circuit provides a pump that creates a voltage drop across a resistor. The voltage drop across the resistor is then related to the current flowing through it by Ohm's law.*

Problem 16.2 This problem is concerned with *units*.

(a) Show the units of inductance L are

$$1\,\mathrm{H} = \frac{1\,\mathrm{Tm}^2}{1\,\mathrm{A}} = \frac{1\,\mathrm{Nm}^2}{1\,\mathrm{Cm/s}}\,\frac{1}{1\,\mathrm{C/s}} = \frac{1\,\mathrm{Nms}^2}{1\,\mathrm{C}^2} \qquad ;\ \text{units of L}$$

(b) Show the units of capacity C are

$$1\,\mathrm{F} = \frac{1\,\mathrm{C}}{1\,\mathrm{V}} = \frac{1\,\mathrm{C}}{1\,\mathrm{J/C}} = \frac{1\,\mathrm{C}^2}{1\,\mathrm{Nm}} \qquad ;\ \text{units of } C$$

(c) Hence, show the units of LC are

$$1\,\mathrm{H} \times 1\,\mathrm{F} = 1\,\mathrm{sec}^2 \qquad ;\ \text{units of L}C$$

Solution to Problem 16.2

(a) We work in SI units. The magnetic field is defined through the Lorentz force law. Thus the unit of magnetic field, the tesla (T) is

$$1\,\mathrm{T} = \frac{1\,\mathrm{N}}{1\,\mathrm{C}\cdot 1\,\mathrm{m/s}}$$

The inductance is a measure of the ratio of magnetic flux to current. Therefore the unit of inductance, the henry (H), is

$$1\,\mathrm{H} = \frac{1\,\mathrm{Tm}^2}{1\,\mathrm{A}} = \frac{1\,\mathrm{Tm}^2}{1\,\mathrm{C/s}}$$

It follows that

$$1\,\mathrm{H} = \frac{1\,\mathrm{Tm}^2}{1\,\mathrm{A}} = \frac{1\,\mathrm{Nm}^2}{1\,\mathrm{Cm/s}}\,\frac{1}{1\,\mathrm{C/s}} = \frac{1\,\mathrm{Nms}^2}{1\,\mathrm{C}^2} \qquad ;\ \text{units of L}$$

(b) The capacity is a measure of the ratio of charge to voltage. Therefore the unit of capacity, the farad (F), is

$$1\,\mathrm{F} = \frac{1\,\mathrm{C}}{1\,\mathrm{V}} = \frac{1\,\mathrm{C}}{1\,\mathrm{J/C}} = \frac{1\,\mathrm{C}^2}{1\,\mathrm{Nm}} \qquad ;\ \text{units of } C$$

(c) A combination of the expressions in parts (a) and (b) then gives

$$1\,\mathrm{H} \times 1\,\mathrm{F} = 1\,\mathrm{sec}^2 \qquad ;\ \text{units of L}C$$

Problem 16.3 (a) Minimize the impedance for the LCR (series) circuit in Eqs. (16.34) and find the resonant angular frequency. What is the minimum impedance? What is the corresponding phase angle?

(b) What are the impedance and phase angle if $R^2 \ll (\omega L - 1/\omega C)^2$?

Solution to Problem 16.3

The expressions in Eqs. (16.34) for the phase angle and impedance (ϕ, Z) characterizing the LRC circuit are[2]

$$\tan\phi = \frac{1}{R}\left(\omega L - \frac{1}{\omega C}\right)$$

$$\left(\frac{\mathcal{E}_0}{i_0}\right)^2 = Z^2 = R^2 + \left(\omega L - \frac{1}{\omega C}\right)^2$$

(a) The square of the impedance increases as the angular frequency ω becomes either very small or very large. The quadratic term in ω is clearly minimized when it vanishes, with

$$\omega = \frac{1}{\sqrt{LC}} \qquad ;\ \text{minimum } Z^2$$

[2]Recall the analysis of the LRC (series) circuit is identical to that of the LCR (series) circuit; we make no distinction between them. (After all, we are discussing oscillations, and the current flows in both directions in the circuit.)

This is the same result obtained for the angular frequency of oscillation in the absence of the resistance R, which introduces dissipation into the circuit. At this resonant frequency, the phase angle vanishes

$$\phi = 0 \qquad ; \; \omega = 1/\sqrt{LC}$$

(b) If $R^2 \ll (\omega L - 1/\omega C)^2$, the R^2 contribution is negligible in Z^2; however, $\tan\phi$ becomes very large, with a sign depending on the relative size of ωL and $1/\omega C$. In this case,

$$\phi \to \pm\frac{\pi}{2} \qquad ; \; R^2 \ll \left(\omega L - \frac{1}{\omega C}\right)^2$$

Problem 16.4 The LRC circuit shown in Fig. 16.1 below is driven by an imperfect frequency generator which puts out with equal amplitudes a desired frequency $\omega_1 = 1\,\text{kHz}$ and undesired ("noise") frequency $\omega_2 = 100\,\text{kHz}$

$$\mathcal{E}(t) = \mathcal{E}_0 \left[\cos(\omega_1 t + \phi_1) + \cos(\omega_2 t + \phi_2)\right]$$

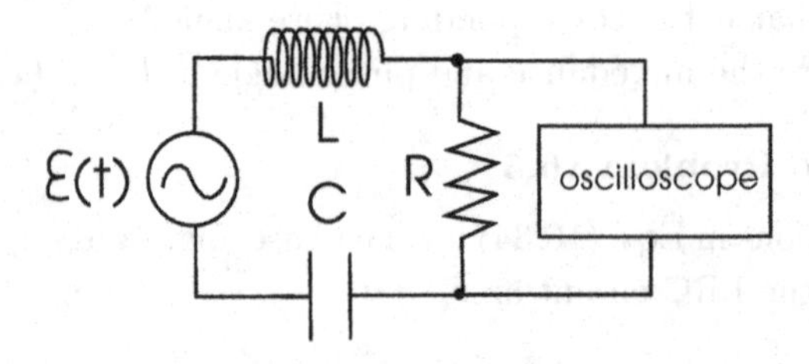

Fig. 16.1 LRC circuit in Prob. 16.4, together with oscilloscope across the resistance.

(a) What would the ratio of $\mathcal{E}_0$ to the peak current be if only the first frequency ω_1 were present? If only the second frequency ω_2?

(b) You have in your laboratory a capacitor with $C = 100\,\mu\text{F}$ and two inductors:

$$L_1 = 10\,\text{mH} \qquad ; \; L_2 = 0.5\,\text{mH}$$

Which combination of capacitor and inductor gives the largest ratio of signal to noise voltage as measured across the resistance R by the oscilloscope?[3]

[3]The LRC (series) circuit then acts as a *band-pass filter*.

Solution to Problem 16.4

Here

$$\omega_1 = 1\,\text{kHz} = 10^3/\text{s} \qquad ; \; \omega_2 = 100\,\text{kHz} = 10^5/\text{s}$$
$$i_1 = i_{10}\cos(\omega_1 t) \qquad ; \; i_2 = i_{20}\cos(\omega_2 t)$$

(a) The *impedance* for an LRC (series) circuit is given in Eqs. (16.34)

$$\frac{\mathcal{E}_0}{i_{10}} = Z_1 = \left[R^2 + \left(\omega_1 L - \frac{1}{\omega_1 C}\right)^2\right]^{1/2}$$
$$\frac{\mathcal{E}_0}{i_{20}} = Z_2 = \left[R^2 + \left(\omega_2 L - \frac{1}{\omega_2 C}\right)^2\right]^{1/2}$$

(b) If both frequencies are present, then by *superposition*

$$i = i_{10}\cos(\omega_1 t) + i_{20}\cos(\omega_2 t)$$

We want to *minimize*[4]

$$\left(\frac{i_{20}}{i_{10}}\right)^2 = \left(\frac{Z_1}{Z_2}\right)^2 = \frac{R^2 + (\omega_1 L - 1/\omega_1 C)^2}{R^2 + (\omega_2 L - 1/\omega_2 C)^2}$$

for different choices of (C, L).

(1) In the first case

$$C_1 = 100\,\mu\text{F} = 10^{-4}\,\text{F} \qquad ; \; L_1 = 10\,\text{mH} = 10^{-2}\,\text{H}$$

In this case, the above ratio is[5]

$$\left(\frac{i_{20}}{i_{10}}\right)^2 = \frac{R^2 + (10 - 1/10^{-1})^2}{R^2 + (10^3 - 1/10)^2} \approx \frac{R^2}{R^2 + 10^6}$$

(2) In the second case

$$C_2 = 100\,\mu\text{F} = 10^{-4}\,\text{F} \qquad ; \; L_2 = \frac{1}{2} \times 10^{-3}\,\text{H}$$

In this case, the above ratio is

$$\left(\frac{i_{20}}{i_{10}}\right)^2 = \frac{R^2 + (1/2 - 1/10^{-1})^2}{R^2 + (10^2/2 - 1/10)^2} \approx \frac{R^2 + 10^2}{R^2 + 10^4/4}$$

The first choice of (C, L) always minimizes $(i_{20}/i_{10})^2$.

[4]It is simpler to deal with the square. Remember, the voltage across the resistance is $\Delta V = iR$.

[5]The units of both the numerator and denominator are Ω^2.

Problem 16.5 The resonant LRC (series) circuit in Fig. 20.7 is analyzed in section 16.5.4 and Probs. 16.3–16.4. The applied EMF $\mathcal{E}$, and current i in the circuit, are given in Eqs. (16.26)

$$\mathcal{E} = \mathcal{E}_0 \cos(\omega t + \phi)$$
$$i = i_0 \cos(\omega t)$$

where ϕ is the phase angle; in general, i and $\mathcal{E}$ are *out of phase.*

(a) Show from Eqs. (16.34) that for this circuit, i_0 and $\mathcal{E}_0$ are related by

$$i_0 R = \frac{\mathcal{E}_0}{(1+\tan^2\phi)^{1/2}} \quad ; \tan\phi = \frac{1}{R}\left(\omega L - \frac{1}{\omega C}\right)$$

(b) Show the voltage signal across the resistor is[6]

$$\Delta V_{\text{signal}} = iR = \frac{\mathcal{E}_0 \cos(\omega t)}{(1+\tan^2\phi)^{1/2}}$$

Discuss the angular frequency dependence of this response.

Solution to Problem 16.5

Figure 20.7 in the text is reproduced here as Fig. 16.2 below.

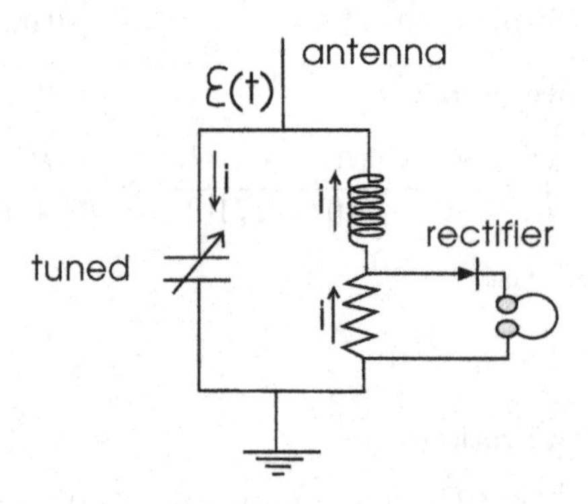

Fig. 16.2 Resonant detector for amplitude modulated (AM) signal.

(a) Equations (16.34) for the phase angle and impedance in the LRC

[6]From Eqs. (16.30), one has $iR = \mathcal{E} - L\,di/dt - q/C$, with $i = dq/dt$. The following problem discusses an alternate means of detecting the signal from the oscillator.

circuit read

$$\tan\phi = \frac{1}{R}\left(\omega L - \frac{1}{\omega C}\right)$$
$$\left(\frac{\mathcal{E}_0}{i_0}\right)^2 = Z^2 = R^2 + \left(\omega L - \frac{1}{\omega C}\right)^2$$

It follows that

$$\frac{Z^2}{R^2} = 1 + \frac{1}{R^2}\left(\omega L - \frac{1}{\omega C}\right)^2 = 1 + \tan^2\phi$$

Hence

$$i_0 R = \frac{\mathcal{E}_0 R}{Z} = \frac{\mathcal{E}_0}{(1+\tan^2\phi)^{1/2}}$$

(b) The voltage signal across the resistor is[7]

$$\Delta V_{\text{signal}} = iR$$

where, from part (a),

$$iR = i_0 R\cos(\omega t) = \frac{\mathcal{E}_0\cos(\omega t)}{(1+\tan^2\phi)^{1/2}}$$

Suppose one is *on resonance*, with $\omega = 1/\sqrt{LC}$. Then

$$\phi = \tan\phi = 0 \qquad ; \ \omega = 1/\sqrt{LC}$$
$$iR = \mathcal{E}_0\cos(\omega t)$$

Here the applied EMF gets passed right through to ΔV_{signal}, and the current in the resonant oscillator becomes very large as the resistance in the circuit becomes small.

On the other hand, suppose one is *off resonance*, with $\tan^2\phi \to \infty$. This condition is

$$\tan^2\phi = \frac{1}{R^2}\left(\omega L - \frac{1}{\omega C}\right)^2 \gg 0$$

As a function of ω, $\tan^2\phi$ has a *minimum* at the resonant value $\omega = 1/\sqrt{LC}$.[8] For small R, $\tan^2\phi$ then grows quickly as ω moves away from

[7]It is assumed the final detector does not affect the oscillator.
[8]See Prob. 16.3.

this resonant value, either increasing or decreasing from it. If the above condition is satisfied, then

$$iR \to 0 \qquad ; \tan^2 \phi \to \infty$$

Off resonance, there is no current in the oscillator, and no voltage signal across the resistor.

Because of the presence of the capacitor and inductor in the oscillating LRC circuit, only the signal at the resonant frequency $\omega = 1/\sqrt{LC}$ gets passed through to the resistor.

Problem 16.6 The signal can also be read out of an oscillator using *mutual inductance*, as discussed in section 14.2.

(a) Suppose one completes the circuit for the second coil in Fig. 14.13 with a resistance R_2. Let the current in the first coil, which sits in the oscillator, be driven as $i_1 = i_{10} \cos(\omega t)$. Show the EMF established in the second coil is then

$$\mathcal{E}_2 = -L_{12}\frac{di_1}{dt} = \omega L_{12} i_{10} \sin(\omega t)$$

where L_{12} is the *mutual inductance* of the two coils.

(b) Suppose a current i_2 now flows in the second circuit. Show Kirchoff's rule for the second circuit gives

$$\mathcal{E}_2 - L_2\frac{di_2}{dt} = i_2 R_2$$

where L_2 is the self-inductance of the second coil.

(c) Write the current in the second coil as $i_2 \equiv i_{20} \cos(\omega t + \eta)$. Show

$$\frac{i_{20}}{i_{10}} = -\frac{L_{12}}{L_2}\frac{1}{(1+\tan^2\eta)^{1/2}} \qquad ; \tan\eta = \frac{R_2}{\omega L_2}$$

The response voltage across the resistor in the second circuit, which can contain a rectifier, is now $\Delta V_{\text{signal}} = i_2 R_2$. Discuss this result.

Solution to Problem 16.6

The configuration for an alternative read-out of the signal from an oscillator using *mutual inductance* is indicated in Fig. 16.3 below.

(a) The magnetic flux through the coil 2 due to the current i_1 in coil 1 is

$$\Phi_2 = L_{12} i_1$$

where L_{12} is the *mutual inductance* of the coils.

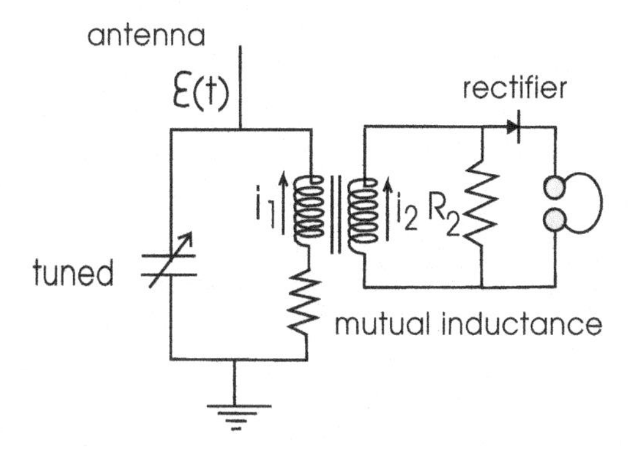

Fig. 16.3 Alternative read-out of an oscillator. The mutual inductance of the two coils, indicated with an appropriate symbol although the coils are most likely co-axial, is L_{12}, while the self-inductance of the second coil is L_2. The current in the oscillator is $i_1 = i_{10}\cos(\omega t)$, while the current in the detecting circuit is $i_2 \equiv i_{20}\cos(\omega t + \eta)$ where $\tan\eta = R_2/\omega L_2$.

It follows from Faraday's law that the EMF generated in coil 2 by the *changing current* in coil 1 is

$$\mathcal{E}_2 = -L_{12}\frac{di_1}{dt} = \omega L_{12} i_{10}\sin(\omega t)$$

(b) Write Kirchoff's second rule for the second circuit[9]

$$\mathcal{E}_2 - L_2\frac{di_2}{dt} = i_2 R$$

Here L_2 is the *self-inductance* of coil 2.

(c) Write the driven current in the second coil as

$$i_2 = i_{20}\cos(\omega t + \eta)$$

It follows that

$$\omega L_{12} i_{10}\sin(\omega t) + \omega L_2 i_{20}\sin(\omega t + \eta) = i_{20} R\cos(\omega t + \eta)$$

[9]Remember that if i goes into a coil, then the increase in EMF as it exits is $-L di/dt$. It is again assumed that the detectors do not affect the oscillators, and for clarity, we have again put the rectifier in the final detection circuit.

Equate coefficients of $\cos(\omega t)$ and $\sin(\omega t)$

$$\omega L_2 i_{20} \sin\eta = i_{20} R \cos\eta$$
$$\omega L_{12} i_{10} + \omega L_2 i_{20} \cos\eta = -i_{20} R \sin\eta$$

Hence

$$\tan\eta = \frac{R}{\omega L_2}$$
$$\frac{i_{20}}{i_{10}} = -\frac{L_{12}}{L_2 \cos\eta + L_2 \tan\eta \sin\eta} = -\frac{L_{12}\cos\eta}{L_2}$$

These equations are re-written as

$$\frac{i_{20}}{i_{10}} = -\frac{L_{12}}{L_2}\frac{1}{(1+\tan^2\eta)^{1/2}} \qquad ; \tan\eta = \frac{R_2}{\omega L_2}$$

The response voltage across the resistor in the second circuit is now $\Delta V_{\text{signal}} = i_2 R_2$

$$\Delta V_{\text{signal}} = i_2 R_2 = i_{20} R_2 \cos(\omega t + \eta)$$

Hence

$$\Delta V_{\text{signal}} = -i_{10}\cos(\omega t + \eta)\,\frac{R_2 L_{12}}{L_2}\frac{1}{(1+\tan^2\eta)^{1/2}}$$

This signal tracks the current in the first oscillator as $i_{10}\cos(\omega t + \eta)$, with a constant of proportionality that depends on $(L_{12}, L_2, R_2, \omega)$.

Problem 16.7 An external signal with an EMF of 1 kilovolt comes into a transformer substation. It is desired to have a useful output of 100 volts. What is the ratio of output yoke-turns to input yoke-turns in the transformer?

Solution to Problem 16.7

This problem is very simple, but it is also informative. If the magnetic flux in the transformer is confined to the iron core, then the ratio of EMF's is just the ratio of the number of turns [see Eq. (16.37)]

$$\frac{\mathcal{E}_1}{\mathcal{E}_2} = \frac{N_1}{N_2} \qquad ; \text{ transformer}$$

Hence

$$\frac{N_{\text{out}}}{N_{\text{in}}} = \frac{\mathcal{E}_{\text{out}}}{\mathcal{E}_{\text{in}}} = \frac{10^2\,\text{V}}{10^3\,\text{V}} = \frac{1}{10}$$

Chapter 17

Review of Magnetism

Problem 17.1 Suppose the particle velocity in Fig. 17.6 acquires a component

$$\vec{v} \rightarrow \vec{v} + \vec{v}_{\perp}$$

where $\vec{v}$ again lies in the plane and $\vec{v}_{\perp}$ is perpendicular to it. Show the only modification of Newton's second law is

$$\frac{d}{dt}(m\vec{v}_{\perp}) = 0$$

Hence conclude that the general particle orbits in this case are spirals along the magnetic field lines.

Solution to Problem 17.1

Let $\vec{v}$ lie in the (x, y)-plane, and $\vec{v}_{\perp}$ and $\vec{B}$ point in the z-direction. Newton's second law then reads

$$\frac{d}{dt}(m\vec{v} + m\vec{v}_{\perp}) = q\,(\vec{v} + \vec{v}_{\perp}) \times \vec{B} = q\,\vec{v} \times \vec{B}$$

If the components are separated, this relation reads

$$\frac{d}{dt}(m\vec{v}) = q\,\vec{v} \times \vec{B}$$
$$\frac{d}{dt}(m\vec{v}_{\perp}) = 0$$

- The first equation for motion in the (x, y)-plane is the one solved in the text;
- The solution to the second equation for motion in the z-direction is $v_z = \text{constant}$;

- Thus, while performing circular motion in the transverse (x, y)-plane, the particle moves with constant velocity in the z-direction;
- *The orbits are spirals about the field lines.*

Problem 17.2 A *thin* toroid of $\mathcal{N}$ closely wound turns of wire of total resistance R lies in the (x, y)-plane. A uniform time-dependent electric field

$$\vec{E} = E_0 \cos \omega t \, \hat{z}$$

is applied in the $\hat{z}$-direction. The inner radius of the toroid is $r = a$ and the circular cross-section of the toroid has radius b. The circuit is initially open (Fig. 17.1 below).

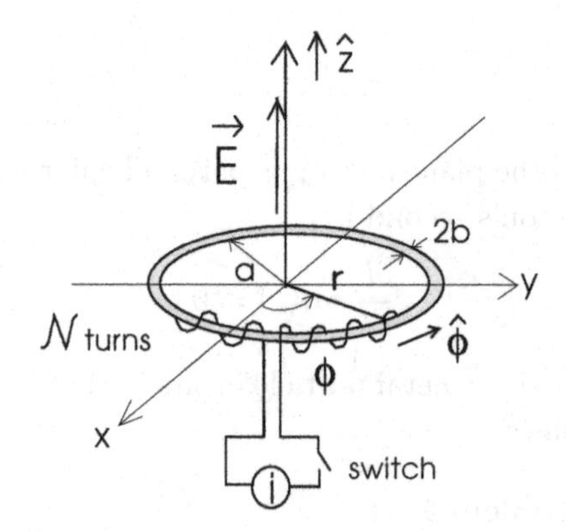

Fig. 17.1 Configuration for Prob. 17.2. The wire wraps completely around the toroid.

(a) Show the magnetic field and magnetic flux through the interior of the toroid are (neglect variations in r)

$$\vec{B} = -\left(\frac{E_0 a \omega}{2c^2} \sin \omega t\right) \hat{\phi}$$

$$\Phi_m = -\mathcal{N}\pi b^2 \left(\frac{E_0 a \omega}{2c^2} \sin \omega t\right)$$

(b) Show that the resulting applied EMF across the open circuit is

$$(\text{EMF})_{\text{app}} = \mathcal{N}\pi b^2 \left(\frac{E_0 a \omega^2}{2c^2} \cos \omega t\right)$$

(c) The switch is now closed. Show that the current is determined by the relation

$$(\text{EMF})_{\text{app}} - L\frac{di}{dt} = iR$$

where $(\text{EMF})_{\text{app}} \equiv V_0 \cos\omega t$, and L is the self-inductance of the toroid.

(d) Determine the amplitude and phase of the current in terms of V_0, L, and R.

Solution to Problem 17.2

(a) The extension of Ampere's law to include the displacement current says that in analogy to Faraday's law, a time-varying electric flux through a surface S induces a magnetic field on the surrounding curve C satisfying [see Eqs. (18.6)–(18.8) and Fig. 18.2][1]

$$\oint_C \vec{B} \cdot d\vec{l} = \frac{1}{c^2}\frac{d}{dt}\int_S \vec{E} \cdot d\vec{S} \qquad ;\ \text{displacement current}$$

In the present situation the curve C is a circle with radius $r \approx a$ (Fig. 17.1 above), and

$$\vec{B} = B_a\,\hat{\phi}$$

It follows from the above relation that

$$2\pi a B_a = \frac{1}{c^2}\frac{d}{dt}\left[\pi a^2 E_0 \cos\omega t\right]$$

Hence

$$B_a = -\left(\frac{E_0 a\omega}{2c^2}\sin\omega t\right)$$

$$\vec{B} = -\left(\frac{E_0 a\omega}{2c^2}\sin\omega t\right)\hat{\phi}$$

The magnetic flux through the coil wrapped around the toroid is then

$$\Phi_m = -\mathcal{N}\pi b^2\left(\frac{E_0 a\omega}{2c^2}\sin\omega t\right)$$

(b) Faraday's law now gives the EMF across the open circuit containing the coil

$$(EMF)_{\text{app}} = -\frac{d\Phi_m}{dt} \qquad ;\ \text{Faraday's law}$$

[1]Unfortunately, this problem got out of order. Part (a) should come after chapter 18 (the problem is actually taken from the course final); however, it is still a good problem, and we will do it here.

Thus

$$(EMF)_{\rm app} = \mathcal{N}\pi b^2 \left(\frac{E_0 a\omega^2}{2c^2}\cos\omega t\right)$$

(c) Kirchoff's second rule applied to the closed circuit then gives

$$(EMF)_{\rm app} - L\frac{di}{dt} = iR$$

where L is the self-inductance of the coil, R is its resistance, and i is the current flowing in it.

(d) From the above, $(EMF)_{\rm app}$ has the time dependence

$$(EMF)_{\rm app} \equiv V_0\cos\omega t \qquad ; \; V_0 = \mathcal{N}\pi b^2 \left(\frac{E_0 a\omega^2}{2c^2}\right)$$

Write the current in the closed circuit as

$$i \equiv i_0 \cos(\omega t + \chi)$$

Substitute these expressions in Kirchoff's rule

$$V_0\cos\omega t + \omega L i_0 \sin(\omega t + \chi) = i_0 R\cos(\omega t + \chi)$$

Now equate the coefficients of $\cos\omega t$ and $\sin\omega t$

$$V_0 + \omega L i_0 \sin\chi = i_0 R \cos\chi$$
$$\omega L i_0 \cos\chi = -i_0 R\sin\chi$$

Hence, the phase of the current is

$$\tan\chi = -\frac{\omega L}{R}$$

Furthermore, the magnitude of the current i_0 is determined by

$$i_0 R = \frac{V_0}{\cos\chi + \sin\chi\tan\chi} = V_0\cos\chi$$

This can be re-written as

$$i_0 R = \frac{V_0}{(1+\tan^2\chi)^{1/2}}$$

The ratio V_0/i_0 defines the effective impedance of the circuit.

PART 3
Electromagnetism

Chapter 18

Maxwell's Equations

Problem 18.1 Derive the integral form of Maxwell's equations from the differential form.

Solution to Problem 18.1

We just work backwards. The differential form of Maxwell's equations in Eqs. (18.22) is[1]

$$\begin{aligned}
\vec{\nabla}\cdot\vec{E}(\vec{x},t) &= \frac{1}{\varepsilon_0}\rho(\vec{x},t) && \text{; Maxwell's equations}\\
\vec{\nabla}\cdot\vec{B}(\vec{x},t) &= 0\\
\vec{\nabla}\times\vec{E}(\vec{x},t) &= -\frac{\partial\vec{B}(\vec{x},t)}{\partial t}\\
\vec{\nabla}\times\vec{B}(\vec{x},t) &= \mu_0\,\vec{j}(\vec{x},t)+\mu_0\varepsilon_0\,\frac{\partial\vec{E}(\vec{x},t)}{\partial t}
\end{aligned}$$

Suppose S is a surface enclosing a volume V as in Fig. 23.1. Then at any instant, the integral over the volume V of the first two equations gives

$$\begin{aligned}
\int_V dv\,[\vec{\nabla}\cdot\vec{E}(\vec{x},t)] &= \int_V dv\,\frac{1}{\varepsilon_0}\rho(\vec{x},t)\\
\int_V dv\,[\vec{\nabla}\cdot\vec{B}(\vec{x},t)] &= 0
\end{aligned}$$

The use of Gauss' theorem on the l.h.s. of these relations then gives Gauss'

[1]Recall that we are working with charges and currents in vacuum.

law for the electric and magnetic fields

$$\int_{\text{closed surface S}} \vec{E} \cdot d\vec{S} = \frac{1}{\varepsilon_0} \int_{\text{enclosed volume V}} \rho\, dv \qquad ; \text{Gauss's law}$$

$$\int_{\text{closed surface S}} \vec{B} \cdot d\vec{S} = 0$$

Suppose C is a curve surrounding a surface S as in Fig. 23.4. Then at any instant, the integral over S of the last two Maxwell equations gives

$$\int_S \left(\vec{\nabla} \times \vec{E}\right) \cdot d\vec{S} = -\int_S \left(\frac{\partial \vec{B}}{\partial t}\right) \cdot d\vec{S}$$

$$\int_S \left(\vec{\nabla} \times \vec{B}\right) \cdot d\vec{S} = \mu_0 \int_S \left[\vec{j} + \varepsilon_0 \frac{\partial \vec{E}}{\partial t}\right] \cdot d\vec{S}$$

Now use Stokes' theorem on the l.h.s. of these relations to derive Faraday's law and Ampere's law with the displacement current.[2]

$$\oint_{\text{closed curve C}} \vec{E} \cdot d\vec{l} = -\frac{d}{dt} \int_{\text{enclosed surface S}} \vec{B} \cdot d\vec{S} \qquad ; \text{Faraday's law}$$

$$\oint_{\text{closed curve C}} \vec{B} \cdot d\vec{l} = \mu_0 \left[\int_{\text{enclosed surface S}} \vec{j} \cdot d\vec{S} + \varepsilon_0 \frac{d}{dt} \int_{\text{enclosed surface S}} \vec{E} \cdot d\vec{S}\right] \qquad ; \text{Ampere's law}$$

Problem 18.2[3] (a) Use the conservation of charge to justify the following relation

$$\int_{\text{closed surface S}} \vec{j}(\vec{x}, t) \cdot d\vec{S} = -\frac{d}{dt} \int_{\text{enclosed V}} \rho(\vec{x}, t) dv$$

Now use Gauss' theorem to deduce the *continuity equation*

$$\vec{\nabla} \cdot \vec{j}(\vec{x}, t) = -\frac{\partial \rho(\vec{x}, t)}{\partial t} \qquad ; \text{continuity equation}$$

(b) Show that for an arbitrary vector field

$$\vec{\nabla} \cdot \left(\vec{\nabla} \times \vec{v}\right) = 0$$

[2]Note Eq. (18.13).

[3]Problems 18.2 and 20.1 are particularly meaningful.

Hence, demonstrate that Maxwell's Eqs. (18.22) imply the continuity equation is satisfied for the source.[4] Notice that it was essential to include the displacement current for this to hold.

Solution to Problem 18.2

(a) It is shown in the text that the charge flowing through the surface $d\vec{S}$ in the time dt is $(\vec{j}\cdot d\vec{S})\,dt$. Hence, if the surface S encloses the volume V, the total charge flowing out through that surface is

$$\text{charge out} = dt \int_{\text{closed surface S}} \vec{j}\cdot d\vec{S}$$

Since charge is conserved, this must be the corresponding *decrease* in the charge contained in V

$$\text{charge out} = -d \int_{\text{enclosed volume V}} \rho\, dV$$

If these two expressions are equated, and the result divided by dt, one finds

$$\int_{\text{closed surface S}} \vec{j}(\vec{x},t)\cdot d\vec{S} = -\frac{d}{dt}\int_{\text{enclosed V}} \rho(\vec{x},t)dv$$

The time derivative can now be moved into the integral on the r.h.s., and Gauss' theorem can be used on the l.h.s., to give

$$\int_{\text{enclosed volume V}} \vec{\nabla}\cdot\vec{j}(\vec{x},t)\,dv = -\int_{\text{enclosed V}} \frac{\partial\rho(\vec{x},t)}{\partial t}dv$$

Since this holds for an arbitrary volume V, one obtains the *continuity equation* for the electromagnetic current

$$\vec{\nabla}\cdot\vec{j}(\vec{x},t) = -\frac{\partial\rho(\vec{x},t)}{\partial t} \qquad ;\ \text{continuity equation}$$

(b) Maxwell's Eqs. (18.22) read

$$\begin{aligned}
\vec{\nabla}\cdot\vec{E}(\vec{x},t) &= \frac{1}{\varepsilon_0}\rho(\vec{x},t) \qquad ;\ \text{Maxwell's equations}\\
\vec{\nabla}\cdot\vec{B}(\vec{x},t) &= 0\\
\vec{\nabla}\times\vec{E}(\vec{x},t) &= -\frac{\partial\vec{B}(\vec{x},t)}{\partial t}\\
\vec{\nabla}\times\vec{B}(\vec{x},t) &= \mu_0\,\vec{j}(\vec{x},t) + \mu_0\varepsilon_0\,\frac{\partial\vec{E}(\vec{x},t)}{\partial t}
\end{aligned}$$

[4]Note that the order of partial derivatives can always be interchanged.

From Prob. 11.1, the divergence of the curl of an arbitrary vector field is obtained as

$$\vec{\nabla}\cdot\left(\vec{\nabla}\times\vec{v}\right)=\varepsilon_{ijk}\left(\frac{\partial}{\partial x_i}\right)\left(\frac{\partial}{\partial x_j}\right)v_k=-\varepsilon_{ijk}\left(\frac{\partial}{\partial x_j}\right)\left(\frac{\partial}{\partial x_i}\right)v_k$$

where we have employed a simple re-labeling of dummy indices ($i \leftrightharpoons j$) and the antisymmetry of ε_{ijk}. This vanishes since the order of the partial derivatives can always be interchanged[5]

$$\vec{\nabla}\cdot\left(\vec{\nabla}\times\vec{v}\right)=0$$

Now take the divergence of the fourth Maxwell equation

$$\vec{\nabla}\cdot\left[\vec{\nabla}\times\vec{B}(\vec{x},t)\right]=\mu_0\left[\vec{\nabla}\cdot\vec{j}(\vec{x},t)+\varepsilon_0\frac{\partial}{\partial t}\vec{\nabla}\cdot\vec{E}(\vec{x},t)\right]=0$$

Substitute the partial time derivative of the first Maxwell equation to obtain

$$\vec{\nabla}\cdot\vec{j}(\vec{x},t)=-\frac{\partial\rho(\vec{x},t)}{\partial t} \qquad ;\ \text{continuity equation}$$

This reproduces the continuity equation for the source, which is here the electromagnetic current in vacuum. Notice that it was essential to include the displacement current in Maxwell's equations for this to hold.

Problem 18.3 Four of the eight Maxwell's Eqs. (18.22) do not involve the sources $(\rho,\vec{j})$, but only the components of the electromagnetic fields $(\vec{E},\vec{B})$. It is convenient to introduce a *scalar potential* $\Phi(\vec{x},t)$ and *vector potential* $\vec{A}(\vec{x},t)$ that allow these four equations to be satisfied identically.[6] Define

$$\vec{E}(\vec{x},t)\equiv-\vec{\nabla}\Phi(\vec{x},t)-\frac{\partial\vec{A}(\vec{x},t)}{\partial t} \qquad ;\ \text{potentials}$$
$$\vec{B}(\vec{x},t)\equiv\vec{\nabla}\times\vec{A}(\vec{x},t)$$

(a) Show that for an arbitrary scalar field

$$\vec{\nabla}\times\left(\vec{\nabla}\chi\right)=0$$

[5]If $a=-a$, then $2a=0$, and $a=0$.

[6]We first met $\Phi(\vec{x},t)$ as the *electrostatic potential* in chapter 5. Recall that as relations between vector fields there are *four* Maxwell equations, but in terms of the individual components there are *eight*. Any reference to the number should be clear from the context.

(b) Hence, show that the following four Maxwell equations hold

$$\vec{\nabla} \cdot \vec{B}(\vec{x}, t) = 0$$
$$\vec{\nabla} \times \vec{E}(\vec{x}, t) = -\frac{\partial \vec{B}(\vec{x}, t)}{\partial t}$$

for *any* $(\Phi, \vec{A})$ above.[7] The other four Maxwell's equations then relate $(\Phi, \vec{A})$ to the sources.

Solution to Problem 18.3

(a) With the use of the analysis in Prob. 11.1, one has

$$\left[\vec{\nabla} \times \left(\vec{\nabla}\chi\right)\right]_i = \varepsilon_{ijk} \frac{\partial}{\partial x_j} \frac{\partial}{\partial x_k} \chi = -\varepsilon_{ijk} \frac{\partial}{\partial x_j} \frac{\partial}{\partial x_k} \chi$$

where the last relation follows from a re-labeling of dummy indices ($j \leftrightharpoons k$), the antisymmetry of ε_{ijk}, and the fact that the partial derivatives can always be interchanged. Hence

$$\vec{\nabla} \times \left(\vec{\nabla}\chi\right) = 0$$

Note this holds for *any* χ.

(b) Now compute the first relation

$$\vec{\nabla} \cdot \vec{B}(\vec{x}, t) = \vec{\nabla} \cdot \left[\vec{\nabla} \times \vec{A}(\vec{x}, t)\right] = 0$$

where this result follows from Prob. 18.2(b).

The second relation then gives

$$\vec{\nabla} \times \vec{E}(\vec{x}, t) = \vec{\nabla} \times \left[-\vec{\nabla}\Phi(\vec{x}, t) - \frac{\partial \vec{A}(\vec{x}, t)}{\partial t}\right] = \vec{\nabla} \times \left[-\frac{\partial \vec{A}(\vec{x}, t)}{\partial t}\right]$$
$$= -\frac{\partial}{\partial t}\left[\vec{\nabla} \times \vec{A}(\vec{x}, t)\right] = -\frac{\partial \vec{B}(\vec{x}, t)}{\partial t}$$

Note that these two Maxwell's equations, which only involve the components of the fields $(\vec{E}, \vec{B})$ and not the sources, are now satisfied for *any* choice of potentials $(\vec{A}, \Phi)$. There is great deal of flexibility in choosing these potentials (see the following problem), which are now related to the sources by the other pair of Maxwell's equations.[8] This flexibility is of invaluable assistance in problem solving.

[7] Recall Prob. 18.2(b).

[8] We leave it to the dedicated reader to write out the other pair. (See, however, the *Aside* below).

Problem 18.4 One has freedom in choosing the potentials in Prob. 18.3, since the defining relations are invariant under a *gauge transformation.* Let $\Lambda(\vec{x}, t)$ be an arbitrary scalar function. Make the following replacements

$$\begin{aligned} \Phi(\vec{x}, t) &\rightarrow \Phi(\vec{x}, t) - \frac{\partial \Lambda(\vec{x}, t)}{\partial t} \qquad ; \text{ gauge transformation} \\ \vec{A}(\vec{x}, t) &\rightarrow \vec{A}(\vec{x}, t) + \vec{\nabla}\Lambda(\vec{x}, t) \end{aligned}$$

Show the electromagnetic fields $(\vec{E}, \vec{B})$ are unchanged.

Solution to Problem 18.4

Make the indicated replacements. Then

$$\begin{aligned} \vec{E}(\vec{x}, t) &\rightarrow -\vec{\nabla}\left[\Phi(\vec{x}, t) - \frac{\partial \Lambda(\vec{x}, t)}{\partial t}\right] - \frac{\partial}{\partial t}\left[\vec{A}(\vec{x}, t) + \vec{\nabla}\Lambda(\vec{x}, t)\right] \\ \vec{B}(\vec{x}, t) &\rightarrow \vec{\nabla} \times \left[\vec{A}(\vec{x}, t) + \vec{\nabla}\Lambda(\vec{x}, t)\right] \end{aligned}$$

Now observe that

- The second and fourth terms in the first line *cancel* since the order of the partial derivatives can always be interchanged;
- The last term in the second line *vanishes* by Prob. 18.3(a).

Hence the fields $(\vec{E}, \vec{B})$ are unchanged

$$\begin{aligned} \vec{E}(\vec{x}, t) &= -\vec{\nabla}\Phi(\vec{x}, t) - \frac{\partial \vec{A}(\vec{x}, t)}{\partial t} \qquad ; \text{ unchanged} \\ \vec{B}(\vec{x}, t) &= \vec{\nabla} \times \vec{A}(\vec{x}, t) \end{aligned}$$

****

(*Aside*) As long as we are at it, we might as well write the relations between the potentials and the sources. The first of Maxwell's equations is

$$\vec{\nabla} \cdot \vec{E} = \frac{1}{\varepsilon_0}\rho \qquad ; \text{ Maxwell's Eqn (1)}$$

Substitute the expression for $\vec{E}$ in terms of the potentials

$$\vec{\nabla} \cdot \left[-\vec{\nabla}\Phi - \frac{\partial \vec{A}}{\partial t}\right] = \frac{1}{\varepsilon_0}\rho$$

This gives

$$\nabla^2\Phi + \frac{\partial}{\partial t}(\vec{\nabla}\cdot\vec{A}) = -\frac{1}{\varepsilon_0}\rho \qquad ; \text{ Maxwell's Eqn (1)}$$

The last Maxwell equation is

$$\vec{\nabla}\times\vec{B} = \mu_0\vec{j} + \frac{1}{c^2}\frac{\partial\vec{E}}{\partial t} \qquad ; \text{ Maxwell's Eqn (4)}$$

Substitute the expression for $(\vec{B},\vec{E})$ in terms of the potentials

$$\vec{\nabla}\times\left[\vec{\nabla}\times\vec{A}\right] = \mu_0\vec{j} + \frac{1}{c^2}\frac{\partial}{\partial t}\left[-\vec{\nabla}\Phi - \frac{\partial\vec{A}}{\partial t}\right]$$

Use Prob. 11.1 to work out the vector relation

$$\begin{aligned}\left\{\vec{\nabla}\times\left[\vec{\nabla}\times\vec{v}\right]\right\}_i &= \varepsilon_{ijk}\varepsilon_{klm}\frac{\partial}{\partial x_j}\frac{\partial}{\partial x_l}v_m \\ &= \left(\delta_{il}\delta_{jm} - \delta_{im}\delta_{jl}\right)\frac{\partial}{\partial x_j}\frac{\partial}{\partial x_l}v_m \\ &= \frac{\partial}{\partial x_i}(\vec{\nabla}\cdot\vec{v}) - \nabla^2 v_i\end{aligned}$$

In vector form this reads[9]

$$\vec{\nabla}\times\left[\vec{\nabla}\times\vec{v}\right] = \vec{\nabla}\,(\vec{\nabla}\cdot\vec{v}) - \nabla^2\,\vec{v}$$

The last Maxwell equation then becomes

$$\left(\nabla^2 - \frac{1}{c^2}\frac{\partial^2}{\partial t^2}\right)\vec{A} - \vec{\nabla}\left[\vec{\nabla}\cdot\vec{A} + \frac{1}{c^2}\frac{\partial\Phi}{\partial t}\right] = -\mu_0\,\vec{j} \qquad ; \text{ Maxwell's Eqn (4)}$$

Gauges: Two commonly employed *gauges*, justified through the freedom provided by gauge transformations, are the following:

(1) *Lorentz Gauge*: The Lorentz gauge is defined through

$$\vec{\nabla}\cdot\vec{A} + \frac{1}{c^2}\frac{\partial\Phi}{\partial t} = 0 \qquad ; \text{ Lorentz gauge}$$

In this gauge, the two Maxwell equations relating the potentials to the sources become

$$\begin{aligned}\Box\Phi &= -\frac{1}{\varepsilon_0}\rho \\ \Box\vec{A} &= -\mu_0\vec{j}\end{aligned}$$

[9]In three dimensions, the laplacian is $\nabla^2 = \partial^2/\partial x^2 + \partial^2/\partial y^2 + \partial^2/\partial z^2$.

where $\Box$ is the wave operator

$$\Box \equiv \nabla^2 - \frac{1}{c^2}\frac{\partial^2}{\partial t^2} \qquad ; \text{ wave operator}$$

In the Lorentz gauge, both potentials $(\Phi, \vec{A})$ satisfy the inhomogeneous wave equation.

(2) *Coulomb gauge*: The Coulomb gauge is defined through[10]

$$\vec{\nabla} \cdot \vec{A} = 0 \qquad ; \text{ Coulomb gauge}$$

In this gauge, the two Maxwell equations relating the potentials to the sources become

$$\nabla^2\Phi = -\frac{1}{\varepsilon_0}\,\rho$$

$$\Box\vec{A} = -\mu_0\,\vec{j}^{\,T} \qquad ; \vec{j}^{\,T} \equiv \vec{j} - \varepsilon_0\vec{\nabla}\frac{\partial\Phi}{\partial t}$$

The continuity equation for the sources implies that the transverse current, as defined here, has no divergence

$$\vec{\nabla} \cdot \vec{j}^{\,T} = \vec{\nabla} \cdot \vec{j} - \varepsilon_0\frac{\partial}{\partial t}\nabla^2\Phi = \vec{\nabla} \cdot \vec{j} + \frac{\partial\rho}{\partial t} = 0$$

One advantage of the Coulomb gauge is that its definition, and the first Maxwell equation, do not explicitly involve the time.

****

Problem 18.5 (a) Write Maxwell's equations in integral and differential form in the *static* case, where everything is independent of time;

(b) Relate these expressions to our previous discussion of *electrostatics* and *magnetostatics*.[11]

Solution to Problem 18.5

(a) Let us refer back to Prob. 18.1. If there is no time dependence in

[10] There is a theorem, proven, for example, in section 64 of [Fetter and Walecka (2003)], that any vector field can be uniquely separated into the sum of a longitudinal and transverse part, with vanishing curl and vanishing divergence, respectively. In the Coulomb gauge, $\vec{A} \equiv \vec{A}^T$.

[11] It is enough here for magnetostatics to obtain Ampere's law. With the aid of these *Problems*, the dedicated reader can also obtain the Biot-Savart law from Maxwell's equations, but that takes some effort.

the system then Maxwell's equations in differential form read

$$\begin{aligned}
\vec{\nabla}\cdot\vec{E}(\vec{x}) &= \frac{1}{\varepsilon_0}\rho(\vec{x}) \\
\vec{\nabla}\cdot\vec{B}(\vec{x}) &= 0 \\
\vec{\nabla}\times\vec{E}(\vec{x}) &= 0 \\
\vec{\nabla}\times\vec{B}(\vec{x}) &= \mu_0\,\vec{j}(\vec{x})
\end{aligned}$$

In integral form they read

$$\begin{aligned}
\int_{\text{closed surface S}} \vec{E}\cdot d\vec{S} &= \frac{1}{\varepsilon_0}\int_{\text{enclosed volume V}} \rho\, dv \\
\int_{\text{closed surface S}} \vec{B}\cdot d\vec{S} &= 0 \\
\oint_{\text{closed curve C}} \vec{E}\cdot d\vec{l} &= 0 \\
\oint_{\text{closed curve C}} \vec{B}\cdot d\vec{l} &= \mu_0 \int_{\text{enclosed surface S}} \vec{j}\cdot d\vec{S}
\end{aligned}$$

(b) The first and third equations just present Gauss' law and the conservative nature of the electric field, which was the basis for our discussion of *electrostatics and DC currents.*

The second and fourth equations present the absence of a magnetic charge and Ampere's law, which provided the basis for our discussion of *magnetostatics.*

Problem 18.6 Show that in electrostatics, the electrostatic potential in a source-free region in two dimensions satisfies Laplace's equation[12]

$$\nabla^2 V \equiv \left(\frac{\partial^2}{\partial x^2} + \frac{\partial^2}{\partial y^2}\right) V = 0 \qquad ;\ \text{Laplace's eqn}$$

Solution to Problem 18.6

In two dimensions, the gradient of a scalar function $V(x,y)$ is

$$\vec{\nabla}V(x,y) = \hat{x}\frac{\partial V}{\partial x} + \hat{y}\frac{\partial V}{\partial y}$$

[12] *Hint:* First establish the vector identity $\vec{\nabla}\cdot\vec{\nabla} = \nabla^2 = \partial^2/\partial x^2 + \partial^2/\partial y^2$, where ∇^2 is the *laplacian.* (Note there is a misprint in the text.)

The divergence of this gradient is then

$$\vec{\nabla} \cdot (\vec{\nabla} V) = \frac{\partial}{\partial x}\left(\frac{\partial V}{\partial x}\right) + \frac{\partial}{\partial y}\left(\frac{\partial V}{\partial y}\right)$$

This can be written as

$$\vec{\nabla} \cdot (\vec{\nabla} V) = \nabla^2 V$$

$$\nabla^2 \equiv \frac{\partial^2}{\partial x^2} + \frac{\partial^2}{\partial y^2} \qquad ; \text{ laplacian}$$

Here ∇^2 is the *laplacian.*

Now in a source-free region in electrostatics, the first Maxwell equation states that

$$\vec{\nabla} \cdot \vec{E} = 0$$

The electric field is related to the potential by $\vec{E} = -\vec{\nabla} V$. Hence, in electrostatics, the electrostatic potential in a source-free region in two dimensions satisfies Laplace's equation

$$\nabla^2 V \equiv \left(\frac{\partial^2}{\partial x^2} + \frac{\partial^2}{\partial y^2}\right) V = 0 \qquad ; \text{ Laplace's eqn}$$

Chapter 19

Waves

Problem 19.1 (a) Show the normal-mode *eigenfunctions* in section 19.4 satisfy the orthonormality relation

$$\frac{2}{l}\int_0^l dx \sin\left(\frac{n\pi x}{l}\right)\sin\left(\frac{m\pi x}{l}\right) = \delta_{m,n}$$

(b) Use this to construct the general solution in Eq. (19.15) for the string with fixed endpoints corresponding to the *initial conditions*[1]

$$\phi(x,0) = f(x)$$
$$\left[\frac{\partial\phi(x,t)}{\partial t}\right]_{t=0} = g(x)$$

Solution to Problem 19.1

(a) Consider the integral $I(m,n)$

$$I(m,n) \equiv \frac{2}{l}\int_0^l dx \sin\left(\frac{n\pi x}{l}\right)\sin\left(\frac{m\pi x}{l}\right)$$

where (m,n) are two positive integers. Use

$$\sin(a)\sin(b) = \frac{1}{2}\left[\cos(a-b) - \cos(a+b)\right]$$

Hence

$$I(m,n) = \frac{1}{l}\int_0^l dx \left\{\cos\left[\frac{(n-m)\pi x}{l}\right] - \cos\left[\frac{(n+m)\pi x}{l}\right]\right\}$$

[1]The wave equation is of second-order in time. Thus we must specify the function and its first time derivative everywhere, at the initial time, to define the solution.

Do the integrals

$$I(m,n) = \left\{ \frac{1}{(n-m)\pi} \sin\left[\frac{(n-m)\pi x}{l}\right] - \frac{1}{(n+m)\pi} \sin\left[\frac{(n+m)\pi x}{l}\right] \right\}_0^l$$

Now observe

- If $m \neq n$, this expression vanishes;
- If $m = n$, the integral is 1.[2]

It follows that

$$I(m,n) = \delta_{m,n}$$

where $\delta_{m,n}$ is the Kronecker delta.

(b) The general solution for the motion of the string with fixed endpoints is

$$\phi(x,t) = \sum_{n=1}^{\infty} A_n \sin\left(\frac{n\pi x}{l}\right) \cos\left[\left(\frac{\pi nct}{l}\right) + \eta_n\right] \qquad ; \text{ general solution}$$

where the amplitudes and phases of the normal mode (A_n, η_n) are constants to be determined from the initial conditions. As stated, we must specify the function and its first time derivative everywhere, at the initial time, to define the solution.

$$\phi(x,0) = f(x) = \sum_{n=1}^{\infty} A_n \sin\left(\frac{n\pi x}{l}\right) \cos(\eta_n)$$

$$\left[\frac{\partial \phi(x,t)}{\partial t}\right]_{t=0} = g(x) = -\sum_{n=1}^{\infty} \left(\frac{\pi nc}{l}\right) A_n \sin\left(\frac{n\pi x}{l}\right) \sin(\eta_n)$$

Now make use of the orthonormality of the eigenfunctions established in part (a)

$$\frac{2}{l}\int_0^l dx\, f(x) \sin\left(\frac{m\pi x}{l}\right) = A_m \cos(\eta_m)$$

$$\frac{2}{l}\int_0^l dx\, g(x) \sin\left(\frac{m\pi x}{l}\right) = -\left(\frac{\pi mc}{l}\right) A_m \sin(\eta_m)$$

This solves for (A_m, η_m) for all m.[3]

---

[2]Start from $\sin^2(a) = [1 - \cos(2a)]/2$.

[3]Take the ratio for $\tan(\eta_m)$, and then use either one for A_m.

Problem 19.2 Show the standing-wave solution in Eq. (19.11) can be interpreted as the superposition of two travelling waves moving in opposite directions[4]

$$\phi(x,t) = \frac{A}{2}\left[\sin k(x-ct) + \sin k(x+ct)\right]$$

Sketch this result.

Solution to Problem 19.2

Use

$$\sin(a+b) = \sin(a)\cos(b) + \cos(a)\sin(b)$$
$$\sin(a-b) = \sin(a)\cos(b) - \cos(a)\sin(b)$$

Hence the superposition of the two running waves gives the standing wave

$$\phi(x,t) = A\sin(kx)\cos(\omega t) \qquad ; \; \omega = kc$$

A sketch of this result is shown in Fig. 19.1 below.

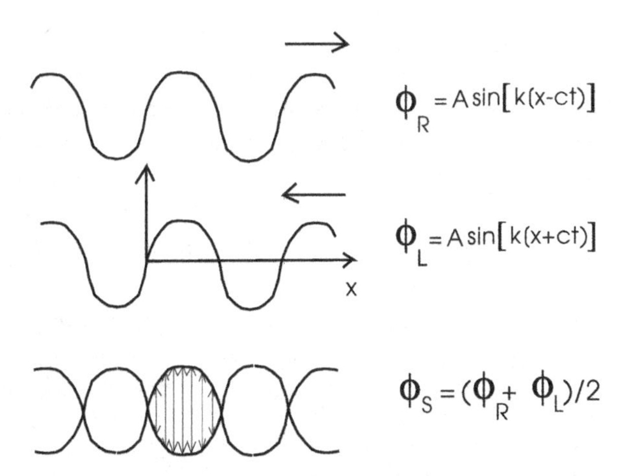

Fig. 19.1 Superposition of the two travelling waves to give a standing wave. The travelling waves are shown at $t = 0$. The waves ϕ_R and ϕ_L then move to the right and left with velocity c.

Note the standing wave satisfies fixed boundary conditions at the nodes, which occur at distances $x = 0$ and $x = d$, where $kd = n\pi$ with $n = 1, 2, \cdots$.

[4]My dictionary says I have the option of putting one or two l's in "travelling"; I have chosen to put two.

The angular frequency of the standing waves is $\omega = kc$.

Problem 19.3 If the far end of the string is *free* in Fig. 19.4, the second boundary condition becomes $[\partial\phi(x,t)/\partial x]_{x=l} = 0$. What are the normal-mode frequencies in this case?

Solution to Problem 19.3

The normal-mode solutions are of the form

$$\phi(x,t) = \phi(x)\cos\omega t$$

In the case of the free end, the spatial boundary conditions are

$$\phi(0) = \left[\frac{d\phi}{dx}\right]_{x=l} = 0$$

It follows that the (unnormalized) eigenfunctions and eigenfrequencies are

$$\phi(x) = \sin\left[\frac{(2n+1)\pi x}{2l}\right] \qquad ; \; n = 0, 1, 2\cdots$$
$$\omega_n = \frac{(2n+1)\pi c}{2l}$$

Problem 19.4 Suppose $f(x)$ represents a plane-wave disturbance on a two-dimensional surface, perhaps a membrane, where the disturbance is independent of y at a given x. Let $\vec{x} = x\,\hat{x} + y\,\hat{y}$ denote an arbitrary position on that surface and let $\hat{n}$ be a unit-vector specifying some direction.

(a) Show $f(\vec{x}\cdot\hat{n} - ct)$ then describe a disturbance that moves without change in shape in the direction $\hat{n}$ with velocity c;

(b) Show this f obeys the *two-dimensional* wave equation[5]

$$\left(\frac{\partial^2}{\partial x^2} + \frac{\partial^2}{\partial y^2}\right) f = \frac{1}{c^2}\frac{\partial^2 f}{\partial t^2} \qquad ; \text{ two dimensions}$$

Solution to Problem 19.4

(a) Suppose one simply rotates the underlying two-dimensional cartesian coordinate system so that the x-axis lies along $\hat{n}$. Then $f(\vec{x}\cdot\hat{n} - ct) = f(x - ct)$, and the discussion of the propagation of the wave is exactly the same as in section 19.1 and Fig. 19.1.

[5] My apologies: the same misprint again appears in the text. I sincerely hope that this (now obvious!) misprint does not detract from this pair of nice two-dimensional problems.

(b) With the un-rotated coordinate system

$$f(\vec{x} \cdot \hat{n} - ct) = f(xn_x + yn_y - t)$$

Consider the following spatial derivatives[6]

$$\frac{\partial f}{\partial x} = n_x f' \qquad ; \quad \frac{\partial f}{\partial y} = n_y f'$$

$$\frac{\partial^2 f}{\partial x^2} = n_x^2 f'' \qquad ; \quad \frac{\partial^2 f}{\partial y^2} = n_y^2 f''$$

Since $\hat{n}$ is a unit vector

$$n_x^2 + n_y^2 = 1$$

The time derivatives are the same as in section 19.1

$$\frac{1}{c^2}\frac{\partial^2 f}{\partial t^2} = f''$$

Hence this f obeys the *two-dimensional* wave equation

$$\left(\frac{\partial^2}{\partial x^2} + \frac{\partial^2}{\partial y^2}\right) f = \frac{1}{c^2}\frac{\partial^2 f}{\partial t^2} \qquad ; \text{ two dimensions}$$

[6]Here $f'(u) = df(u)/du$.

Chapter 20

Electromagnetic Waves

Problem 20.1 The energy density and energy flux for the electromagnetic field in vacuum are given in Eqs. (20.18) and (20.24)

$$U = \frac{\varepsilon_0}{2}|\vec{E}\,|^2 + \frac{1}{2\mu_0}|\vec{B}\,|^2 \qquad ; \text{ energy density}$$
$$\vec{S} = \frac{1}{\mu_0}\vec{E}\times\vec{B} \qquad ; \text{ energy flux}$$

(a) Use the analysis in Prob. 11.1 to establish the vector identity

$$\vec{\nabla}\cdot(\vec{a}\times\vec{b}) = \vec{b}\cdot(\vec{\nabla}\times\vec{a}) - \vec{a}\cdot(\vec{\nabla}\times\vec{b})$$

(b) Now use Maxwell's equations in vacuum to establish the *continuity equation for the electromagnetic energy*

$$\frac{\partial U(\vec{x},t)}{\partial t} + \vec{\nabla}\cdot\vec{S}(\vec{x},t) = 0 \qquad ; \text{ continuity equation}$$

This is energy conservation [see Prob. 18.2(a)].

Solution to Problem 20.1

(a) From Prob. 11.1, the divergence of the cross product can be written

$$\left(\frac{\partial}{\partial x_i}\right)\varepsilon_{ijk}a_jb_k = \varepsilon_{ijk}\left(\frac{\partial a_j}{\partial x_i}\right)b_k + \varepsilon_{ijk}a_j\left(\frac{\partial b_k}{\partial x_i}\right)$$

Hence

$$\vec{\nabla}\cdot(\vec{a}\times\vec{b}) = \vec{b}\cdot(\vec{\nabla}\times\vec{a}) - \vec{a}\cdot(\vec{\nabla}\times\vec{b})$$

(b) From Eqs. (20.2), Maxwell's equations in vacuum read

$$\begin{aligned}
\vec{\nabla}\cdot\vec{E}(\vec{x},t) &= 0 &&; \text{ Maxwell's equations}\\
\vec{\nabla}\cdot\vec{B}(\vec{x},t) &= 0\\
\vec{\nabla}\times\vec{E}(\vec{x},t) &= -\frac{\partial\vec{B}(\vec{x},t)}{\partial t}\\
\vec{\nabla}\times\vec{B}(\vec{x},t) &= \mu_0\varepsilon_0\,\frac{\partial\vec{E}(\vec{x},t)}{\partial t}
\end{aligned}$$

Now take the divergence of the Poynting vector

$$\vec{\nabla}\cdot\vec{S} = \frac{1}{\mu_0}\vec{\nabla}\cdot\left(\vec{E}\times\vec{B}\right) = \frac{1}{\mu_0}\left[\vec{B}\cdot(\vec{\nabla}\times\vec{E}) - \vec{E}\cdot(\vec{\nabla}\times\vec{B})\right]$$

Substitute Maxwell's equations

$$\begin{aligned}
\vec{\nabla}\cdot\vec{S} &= -\frac{1}{\mu_0}\vec{B}\cdot\left(\frac{\partial\vec{B}}{\partial t}\right) - \varepsilon_0\vec{E}\cdot\left(\frac{\partial\vec{E}}{\partial t}\right)\\
&= -\frac{\partial}{\partial t}\left[\frac{1}{2\mu_0}(\vec{B}\cdot\vec{B}) + \frac{\varepsilon_0}{2}(\vec{E}\cdot\vec{E})\right] = -\frac{\partial U}{\partial t}
\end{aligned}$$

This is the *continuity equation for the electromagnetic energy density*

$$\frac{\partial U(\vec{x},t)}{\partial t} + \vec{\nabla}\cdot\vec{S}(\vec{x},t) = 0 \qquad ; \text{ continuity equation}$$

This derivation serves as the ultimate justification for our physical interpretation of $(\vec{S}, U)$.

Problem 20.2 Although we leave its development to another course, Eq. (20.40) provides a simple introduction to *optics.* Suppose one has the coherent superposition of two electromagnetic waves that have travelled a different spatial distance d

$$E_y = A\left\{\cos\left[k(x+d) - \omega t\right] + \cos\left[kx - \omega t\right]\right\}$$

Coherence here implies the same (k, ω) and a constant relative amplitude, as might be obtained, say, by passing a plane wave through two slits. Show that for distances $kd = n\pi$, with $n = 0, 1, 2, \cdots$, the *interference* pattern is

$$E_y = A\left[1 + (-1)^n\right]\cos\left[kx - \omega t\right] \qquad ; \; kd = n\pi$$

Interpret this result.

Solution to Problem 20.2

Start from

$$E_y = A\{\cos[k(x+d) - \omega t] + \cos(kx - \omega t)\}$$

Now use

$$\cos(a+b) = \cos(a)\cos(b) - \sin(a)\sin(b)$$

Then

$$\cos[k(x+d) - \omega t] = \cos(kx - \omega t)\cos(kd) - \sin(kx - \omega t)\sin(kd)$$

Suppose the difference in spatial distance travelled satisfies

$$kd = n\pi \qquad ; \; n = 0, 1, 2, \cdots$$

This gives

$$\cos(kd) = (-1)^n \qquad ; \; \sin(kd) = 0$$

Hence for distances $kd = n\pi$, with $n = 0, 1, 2, \cdots$, the *interference* pattern is

$$E_y = A[1 + (-1)^n]\cos(kx - \omega t) \qquad ; \; kd = n\pi$$

This interference condition is

$$kd = \frac{2\pi d}{\lambda} = n\pi$$

which implies that

$$n\frac{\lambda}{2} = d$$

If the distance d is an *even* number of half-wavelengths,[1] one has *constructive* interference. If it is an *odd* number of half-wavelengths, one has *destructive* interference.

[1]That is, an *integer* number of wavelengths.

Chapter 21

More Electromagnetic Waves

Problem 21.1 Show that the small displacement of a string with constant mass density μ under a constant tension τ obeys the one-dimensional wave Eq. (21.1).[1]

Solution to Problem 21.1

Suppose the string is stretched along the x-axis and free to move in the (x, y)-plane. let $\phi(x,t)$ be a small displacement of the string in the y-direction (see Fig. 21.1 in the text), and dx a little element of the string at position x. At any given instant, there is a (constant) tension force τ directed along the string. It pulls the far end of the little element in the positive x-direction, and the rear end in the negative x-direction. Let $\chi(x)$ be the angle the tangent to the string makes with the x-axis. The y-component of the net tension force on the little element at any instant is then

$$\begin{aligned} F_y &= \tau\left[\sin\chi(x+dx) - \sin\chi(x)\right] \approx \tau\left[\tan\chi(x+dx) - \tan\chi(x)\right] \\ &= \tau\left[\left(\frac{\partial\phi}{\partial x}\right)_{x+dx} - \left(\frac{\partial\phi}{\partial x}\right)_x\right] \approx \tau\, dx\left(\frac{\partial^2\phi}{\partial x^2}\right) \end{aligned}$$

The mass times the acceleration in the y-direction of the little element is

$$dm\,\ddot{y} = \mu\, dx\left(\frac{\partial^2\phi}{\partial t^2}\right)$$

Newton's second law then gives

$$\mu\, dx\left(\frac{\partial^2\phi}{\partial t^2}\right) = \tau\, dx\left(\frac{\partial^2\phi}{\partial x^2}\right) \qquad ;\ \text{Newton's second law}$$

[1] See, for example, [Fetter and Walecka (2003)].

The limit $dx \to 0$ produces the one-dimensional wave Eq. (21.1)

$$\frac{\partial^2 \phi}{\partial x^2} = \frac{1}{c^2}\frac{\partial^2 \phi}{\partial t^2} \qquad ; \; c^2 = \frac{\tau}{\mu} \quad , \text{ wave velocity}$$

Problem 21.2 It is desired to tune into the AM broadcast band with a frequency range of $\nu = 1\text{–}2\,\text{MHz}$. An LC oscillator is available with an inductance of $1\,\mu\text{H}$.[2] What range of capacity of the variable capacitor is required, in μF?

Solution to Problem 21.2

From Eq. (16.12), the angular frequency of an LC oscillator is

$$\omega = \frac{1}{\sqrt{LC}}$$

Since the angular frequency is related to the frequency by $\omega = 2\pi\nu$, this relation can be re-written as

$$C = \frac{1}{(2\pi\nu)^2 L}$$

The insertion of the given numbers, and reference to Prob. 16.2, gives the required capacity for the lower frequency as

$$\begin{aligned} C &= \frac{1}{(2\pi \times 10^6/\text{sec})^2}\frac{1}{1 \times 10^{-6}\,\text{H}} \\ &= \frac{1}{(2\pi)^2}\,\mu\text{F} \qquad ; \; \nu = 1\,\text{MHz} \end{aligned}$$

If the frequency is doubled, the required capacity is

$$C = \frac{1}{(4\pi)^2}\,\mu\text{F} \qquad ; \; \nu = 2\,\text{MHz}$$

Problem 21.3 Show that a charged particle moving in the $\hat{y}$-direction at $x = l/2$ with the mode in the cavity in Fig. 21.3 feels an electric field in the $\hat{y}$-direction oscillating with a frequency $\nu = c/2l$, and no magnetic field.[3]

Solution to Problem 21.3

Figure 21.3 in the text is reproduced as Fig. 21.1 below.

---

[2]Here MHz means "megahertz", or 10^6 Hz, while μH means "microhenry", or 10^{-6} H.
[3]Phased arrays of such cavities are used as particle accelerators.

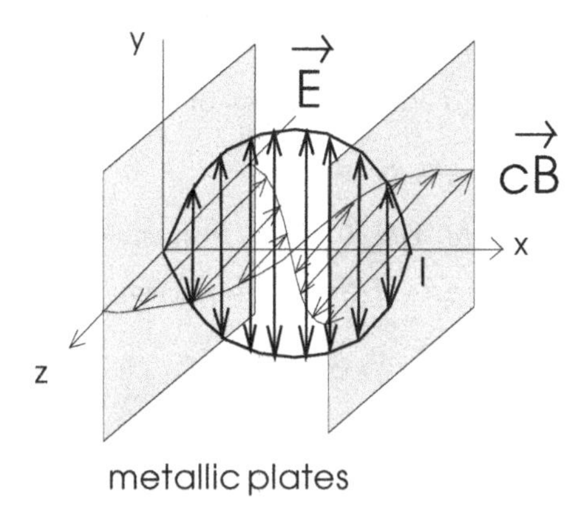

Fig. 21.1 Electromagnetic standing waves between two parallel metal plates perpendicular to the x-axis and separated by a distance l. Shown is the fundamental mode with $n = 1$.

Consider a charged particle moving in the $\hat{y}$-direction at any z in the cavity in Fig. 21.1 above.[4] The Lorentz force is

$$\vec{F} = q\left(\vec{E} + \vec{v} \times \vec{B}\right)$$

If the orbit sits at $x = l/2$, the magnetic field $\vec{B}$ vanishes for all times. The particle then feels no magnetic force, and it moves in a straight line. There is a maximal electric field $\vec{E}$ pointing in the direction of motion (the $\hat{y}$-direction), which oscillates with time with a frequency given by Eq. (21.16)

$$\nu = \frac{nc}{2l}$$

For the fundamental mode in Fig. 21.1 above, $n = 1$.

With a phased array of such cavities arranged so that as the particle moves through a cavity, the electric field and electric force always point in the same direction, one has a particle *accelerator*.

Problem 21.4 The electromagnetic spectrum represents an invaluable resource.

(a) Find out what part of the spectrum is used by today's smartphones and wi-fi devices;

(b) Find out how the electromagnetic spectrum is managed and distributed.

[4]Note the mode is independent of z.

Solution to Problem 21.4

Here are three places to start (courtesy of Google)

https://en.wikipedia.org/wiki/Wi-Fi

https://en.wikipedia.org/wiki/Cellular frequencies in the US

https://en.wikipedia.org/wiki/Spectrum auction

Chapter 22

The Theory of Special Relativity

Problem 22.1 Show that the Lorentz transformation in Eqs. (22.2)–(22.3) leaves the quadratic form in Eq. (22.4) invariant.

Solution to Problem 22.1

The Lorentz transformation in Eqs. (22.3) is

$$x' = \frac{(x - vt)}{\sqrt{1 - v^2/c^2}} \qquad ; \; t' = \frac{(t - vx/c^2)}{\sqrt{1 - v^2/c^2}}$$

Calculate

$$\begin{aligned} x'^2 - c^2t'^2 &= \frac{1}{1 - v^2/c^2} \left[(x - vt)^2 - c^2(t - vx/c^2)^2\right] \\ &= \frac{1}{1 - v^2/c^2} \left[x^2 \left(1 - \frac{v^2}{c^2}\right) - c^2t^2 \left(1 - \frac{v^2}{c^2}\right)\right] \end{aligned}$$

Thus it is demonstrated that the Lorentz transformation leaves this quadratic form invariant

$$x'^2 - c^2t'^2 = x^2 - c^2t^2$$

Problem 22.2 Take the travelling-wave solution to the one-dimensional wave equation in Eq. (20.11), substitute the Lorentz transformation in Eqs. (22.2), and show the new wave satisfies the same wave Eq. (22.10), with the same velocity c.

Solution to Problem 22.2

The travelling-wave in Eq. (20.11) is

$$E_y(x, t) = A \cos\left[k(x - ct)\right]$$

It satisfies the one-dimensional wave Eq. (22.10)

$$\left(\frac{\partial^2}{\partial x^2} - \frac{1}{c^2}\frac{\partial^2}{\partial t^2}\right) E_y(x,t) = 0$$

The Lorentz transformation in Eqs. (22.2) is

$$x \equiv \frac{(x' + vt')}{\sqrt{1 - v^2/c^2}} \qquad ; \text{ Lorentz transformation}$$

$$t \equiv \frac{(t' + vx'/c^2)}{\sqrt{1 - v^2/c^2}}$$

Substitute this Lorentz transformation in the above solution

$$E_y[x(x',t'), t(x',t')] = A\cos\left\{\frac{k}{\sqrt{1 - v^2/c^2}}\left[(x' + vt') - c\left(t' + vx'/c^2\right)\right]\right\}$$

This expression simplifies to

$$E_y(x',t') = A\cos\left[k'\left(x' - ct'\right)\right] \qquad ; \; k' \equiv k\left(\frac{1 - v/c}{1 + v/c}\right)^{1/2}$$

The new wave evidently *satisfies the same wave Eq. (22.10), with the same velocity c*

$$\left(\frac{\partial^2}{\partial x'^2} - \frac{1}{c^2}\frac{\partial^2}{\partial t'^2}\right) E_y(x',t') = 0$$

The only change is the modification ("Doppler shift") of the wavenumber $k \to k'$.

Problem 22.3 (a) A muon lives $\tau_\mu = 2.2 \times 10^{-6}$ sec in its rest frame. How fast must it be moving to have its laboratory lifetime extended to 10^{-3} sec?

(b) The total energy of a particle moving with velocity v in special relativity is

$$E = \frac{m_0 c^2}{\sqrt{1 - v^2/c^2}} \qquad ; \text{ particle energy}$$

where m_0 is the particle's rest mass. The rest mass of the muon is $m_\mu c^2 =$ 105.6 MeV. What is the muon's energy in part (a), in GeV? [1]

(c) The muon in part (a) moves down a tube 100 m long. How long does the tube appear to be if you sit on the muon?

[1] Note 1 MeV = 10^6 eV, and 1 GeV = 10^9 eV.

Solution to Problem 22.3

(a) The time dilation in special relativity gives the laboratory lifetime of the moving muon as

$$t = \frac{\tau_\mu}{\sqrt{1 - v^2/c^2}}$$

Hence

$$10^{-3}\,\text{sec} = \frac{1}{\sqrt{1 - v^2/c^2}}\,2.2 \times 10^{-6}\,\text{sec}$$
$$\sqrt{1 - v^2/c^2} = 2.2 \times 10^{-3}$$

It follows that

$$\frac{v}{c} = \sqrt{1 - 4.84 \times 10^{-6}}$$
$$\approx 1 - 2.42 \times 10^{-6}$$

(b) The energy of the moving muon is

$$E = \frac{m_0 c^2}{\sqrt{1 - v^2/c^2}} = \frac{105.6\,\text{MeV}}{2.2 \times 10^{-3}}$$
$$= 4.8 \times 10^4\,\text{MeV} = 48\,\text{GeV}$$

(c) The Lorentz-contracted length of a 100 m tube is then

$$l = \sqrt{1 - v^2/c^2} \times 10^2\,\text{m} = 0.22\,\text{m}$$
$$= 22\,\text{cm}$$

Problem 22.4 What is the role of the second of the Lorentz transformation Eqs. (22.3) in the discussion of Lorentz contraction in section 22.4.3?

Solution to Problem 22.4

The second of the Lorentz transformation Eqs. (22.3) is

$$t' = \frac{(t - vx/c^2)}{\sqrt{1 - v^2/c^2}}$$

It relates the time t' in the second frame to the time and position (t, x) in the first frame.

In the discussion of Lorentz contraction, the length of the moving meter stick is measured simultaneously $(t = 0)$ at two positions in the first frame. In the second frame (the rest frame of the meter stick), the times at which the position of those two ends is measured will differ; however, since the meter stick is *at rest*, it does not matter at which times those positions are measured – they do not change.

Problem 22.5 A large, square, charged parallel-plate capacitor with a field $\vec{E}$ between the plates is set in motion with a velocity $\vec{v}$ perpendicular to one of its sides, and to $\vec{E}$.

(a) Use Eq. (22.17) to determine the magnetic field $\vec{B}$ between the plates;

(b) Sketch the configuration. Interpret the result in (a) in terms of the currents provided by the plates.[2]

Solution to Problem 22.5

(a) The electric field $\vec{E}$ between the plates of a large, square, charged parallel-plate capacitor is calculated in Eq. (3.17) to be (see also Prob. 4.3)

$$\vec{E} = \frac{\sigma}{\varepsilon_0}\hat{z} \qquad ; \text{ between two sheets with opposite charge}$$

Equation (22.17) from special relativity states that with every moving electric field $\vec{E}$, there is an associated magnetic field $\vec{B}$

$$\vec{B} = \frac{1}{c^2}\vec{v} \times \vec{E} = \mu_0\varepsilon_0\,\vec{v} \times \vec{E}$$

In the present case, this field points in the $\hat{y}$ direction in Fig. 13.5, where the velocity of the plates is in the $-\hat{x}$-direction (see Fig. 22.1 below).

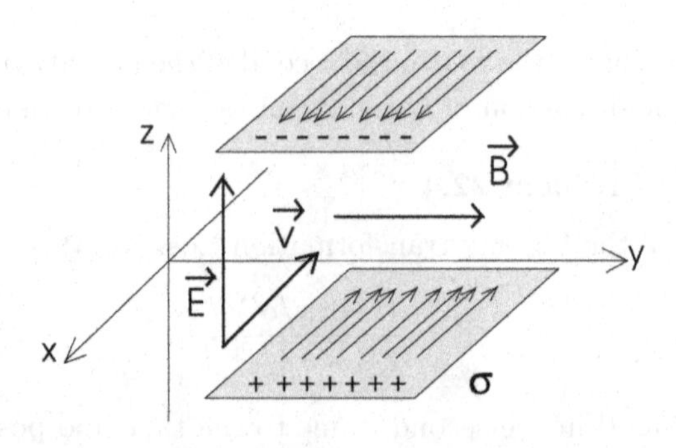

Fig. 22.1 A large, square, charged parallel-plate capacitor with a field $\vec{E} = (\sigma/\varepsilon_0)\hat{z}$ between the plates is set in motion with a velocity $\vec{v}$ perpendicular to one of its sides, and to $\vec{E}$. The arrows on the sheets denote the current.

[2]See the discussion in section 13.2.2.

The magnetic field between the moving plates thus takes the form

$$\vec{B} = \mu_0 \, \sigma v \, \hat{y}$$

The movement of the charged plate gives rise to a *current*. The amount of charge dq crossing unit transverse length in time dt is $\sigma v \, dt$. Hence the current per unit transverse length of the moving plate is

$$\eta \equiv \frac{dq}{dt} = \sigma v \qquad ; \text{ current/length}$$

Hence

$$\vec{B} = \mu_0 \eta \, \hat{y}$$

(b) The charged, moving plates provide two opposed current sheets, and the field between two such sheets is calculated from Ampere's law to be [see Eq. (13.15)]

$$\vec{B} = \mu_o \eta \, \hat{y} \qquad ; \text{ between sheets}$$

This is exactly the result given above.

From this we learn that electrostatics and special relativity can reproduce the results of Ampere's law and magnetostatics, at least in this case.

Problem 22.6 (a) Compute the energy flux $\vec{S}$ as the electric field in Prob. 22.5 moves through space.
(a) Show that to lowest order in v/c

$$\vec{S} = \left(\frac{\varepsilon_0}{2} |\vec{E}\,|^2\right) \vec{v} \qquad ; \vec{v} \to 0$$

(b) Show that as $\vec{v} \to \vec{c}$, the energy flux is given by the Poynting vector[3]

$$\vec{S} = \frac{1}{\mu_0} \vec{E} \times \vec{B} \qquad ; \vec{v} \to \vec{c}$$

Solution to Problem 22.6

We make use of the fact that in the configuration of Prob. 22.5, the velocity $\vec{v}$ of the capacitor, and the electric field $\vec{E}$ in the capacitor, are perpendicular

$$\vec{v} \cdot \vec{E} = 0$$

[3]Recall Prob. 20.1 and the vector identities in Prob. 11.1.

We also make use of the two vector identities from Prob. 11.1

$$(\vec{v} \times \vec{E}) \cdot (\vec{v} \times \vec{E}) = \vec{v}^{\,2}\vec{E}^{\,2} - (\vec{v} \cdot \vec{E})^2 = \vec{v}^{\,2}\vec{E}^{\,2}$$
$$\vec{E} \times (\vec{v} \times \vec{E}) = \vec{v}\,\vec{E}^{\,2} - \vec{E}(\vec{v} \cdot \vec{E}) = \vec{v}\,\vec{E}^{\,2}$$

(a) The energy flux of the electric field is given by the energy density of the field times its velocity

$$S_E = U_E\,\vec{v} = \left(\frac{\varepsilon_0}{2}|\vec{E}\,|^2\right)\vec{v}$$

Equation (22.17) from special relativity states that with every moving electric field $\vec{E}$, there is an associated magnetic field $\vec{B}$

$$\vec{B} = \frac{1}{c^2}\vec{v} \times \vec{E}$$

If the magnetic field moves with $\vec{E}$,[4] the energy flux of the magnetic field is similarly given by the energy density of the field times its velocity

$$S_B = U_B\,\vec{v} = \left(\frac{1}{2\mu_0}|\vec{B}\,|^2\right)\vec{v}$$

Now work out

$$\frac{1}{2\mu_0 c^4}(\vec{v} \times \vec{E}) \cdot (\vec{v} \times \vec{E}) = \frac{\varepsilon_0}{2}\frac{v^2}{c^2}|\vec{E}\,|^2$$

Hence

$$S_B = U_B\,\vec{v} = \frac{v^2}{c^2}\left(\frac{\varepsilon_0}{2}|\vec{E}\,|^2\right)\vec{v}$$

The total energy flux of the fields inside the moving capacitor is therefore

$$S_E + S_B = \frac{1}{2}\left(1 + \frac{v^2}{c^2}\right)\left(\varepsilon_0|\vec{E}\,|^2\right)\vec{v}$$

The magnetic contribution is negligible as $v/c \to 0$. It is equal to the electric contribution as $v/c \to 1$

(b) The above result gives the energy flux arising from the fields $(\vec{E}, \vec{B}\,)$ produced by a moving source. The *Poynting vector* gives the energy flux from self-sustaining fields $(\vec{E}, \vec{B}\,)$ in vacuum (that is, from radiation).[5] The Poynting vector is

$$S = \frac{1}{\mu_0}\vec{E} \times \vec{B}$$

[4]Recall Eqs. (22.15)–(22.16).
[5]Recall Prob. 20.1.

A calculation of this quantity in the present case gives

$$\frac{1}{\mu_0}\vec{E}\times\vec{B} = \varepsilon_0\,\vec{E}\times(\vec{v}\times\vec{E}\,) = \left(\varepsilon_0|\vec{E}\,|^2\right)\vec{v}$$

This is the same result obtained in part (a) as $v/c \to 1$.

Chapter 23

Review of Electromagnetism

Problem 23.1 A radio station in San Francisco emits electromagnetic waves with an average power of $\bar{P} = 10^5\,\mathrm{W}$ (Fig. 23.1 below).

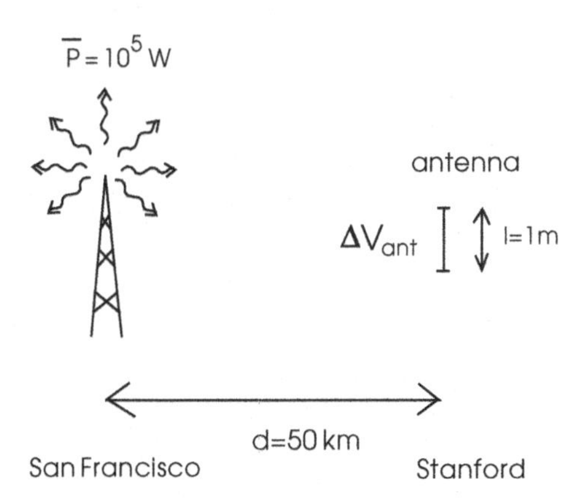

Fig. 23.1 Configuration for Prob. 23.1.

(a) Calculate the time-average Poynting vector $\bar{S}$ at Stanford, a distance of $d = 50\,\mathrm{km}$ away.[1]

(b) From this, calculate the amplitude of the $\vec{E}$ and $\vec{B}$ fields in the radio signal received at Stanford.

(c) A dipole antenna is straight wire of length $l = 1\,\mathrm{m}$ across which one measures the voltage. The electromagnetic wave hits the wire (uniformly) and produces a voltage. What is the maximum voltage ΔV_{ant} produced in the antenna at Stanford?

[1] *Hint*: The electromagnetic radiation is emitted in a spherically symmetric fashion.

Solution to Problem 23.1

(a) The transmitter emits a time-average power $\bar{P} = 10^5\,\text{W}$ (energy/time) isotropically. The time-average Poynting vector $\bar{S}$ (average power per square meter) a distance d away is then given by

$$\bar{P} = 4\pi d^2 \bar{S}$$

Hence

$$\bar{S} = \frac{1}{4\pi}\frac{10^5\,\text{W}}{(50\times 10^3\,\text{m})^2} = \frac{1}{4\pi}\frac{10^5\,\text{W}}{25\times 10^8\,\text{m}^2}$$

(b) The cosine wave of Eqs. (20.11)–(20.12) is

$$E_y(x,t) = cB_z(x,t) = A\cos\left[k(x-ct)\right]$$

In terms of the fields, the magnitude of the time-average Poynting vector is then[2]

$$|\bar{S}| = \frac{1}{\mu_0}\langle\,|\vec{E}\times\vec{B}|\,\rangle = \frac{1}{\mu_0 c}A^2\langle\,\cos^2\left[k(x-ct)\right]\,\rangle = \frac{1}{2\mu_0 c}A^2$$

Hence, the amplitude A of the electric field $\vec{E}$ in the wave at the antenna is given by

$$\begin{aligned} A^2 &= 2\mu_0 c\,\bar{S} = \frac{\mu_0}{4\pi}\left(2\times 3\times 10^8\,\frac{\text{m}}{\text{s}}\right)\frac{10^5\,\text{W}}{25\times 10^8\,\text{m}^2} \\ &= 10^{-7}\,\frac{\text{Ns}^2}{\text{C}^2}\left(6\times 10^8\,\frac{\text{m}}{\text{s}}\right)\frac{10^5\,\text{W}}{25\times 10^8\,\text{m}^2} \end{aligned}$$

Make use of the following unit conversions

$$\begin{aligned} 1\,\text{Nm} &= 1\,\text{VC} \\ 1\,\text{W} &= 1\,\text{VC/s} \end{aligned}$$

It follows that

$$A^2 = \frac{6\times 10^{-2}}{25}\,\frac{(\text{VC/m})\,\text{s}^2}{\text{C}^2}\,\frac{\text{m}}{\text{s}}\,\frac{\text{VC/s}}{\text{m}^2} = \frac{6\times 10^{-2}}{25}\left(\frac{\text{V}}{\text{m}}\right)^2$$

Hence

$$A = \frac{\sqrt{6}}{50}\,\frac{\text{V}}{\text{m}}$$

[2]The notation $\langle\cdots\rangle$ denotes a time average. Here we confine the analysis to a single polarization.

(c) The voltage difference is related to the electric field by

$$\Delta V = -\int_1^2 \vec{E}\cdot d\vec{l}$$

Thus the maximum voltage across the receiving antenna is

$$|\Delta V_{\rm ant}|_{\rm max} = |\vec{E}|_{\rm max}\, l = Al \qquad ; \text{ maximum voltage}$$

Hence, for $l = 1\,\text{m}$,

$$|\Delta V_{\rm ant}|_{\rm max} = \frac{\sqrt{6}}{50}\,\text{V}\cdot$$

Problem 23.2 The device in Prob. 4.6 is viewed from a frame moving with a velocity $-\vec{v}$ pointing along the charge line ($v/c \ll 1$). Work in that frame.

(a) Show that the current arising from the moving line of charge is $i = \lambda v$;

(b) Use Ampere's law to show that for $0 < |\vec{\rho}| < a$ there is an azimuthal magnetic field

$$\vec{B} = \frac{\mu_0 i}{2\pi\rho}\,\hat{\phi} \qquad ; \rho \equiv |\vec{\rho}|$$

(c) What is the azimuthal magnetic field for $a < |\vec{\rho}| < b$?

(d) What is the magnetic field for $|\vec{\rho}| > b$?

(e) Use Eq. (22.17), and derive the above results from the electric fields in Prob. 4.6.

Solution to Problem 23.2

(a) The device in Prob. 4.6 is a line of charge surrounded by a conducting cylinder. From the frame moving with $-\vec{v}$ in the direction of the line, the line of charge appears to be a *current*. If λ is the charge per unit length in the line, then the amount of charge dq moving past a point in the time dt is $\lambda v\, dt$. Hence the observed current is

$$i = \frac{dq}{dt} = \lambda v \qquad ; \text{ current}$$

(b) Surround the line of current with a transverse amperian loop whose radius is less than the inside radius of the conducting cylinder. The current through this loop is i, and there will be a magnetic field surrounding this current of the form

$$\vec{B} = B_\rho\,\hat{\phi}$$

where ϕ is the azimuthal angle. Ampere's law then gives

$$\oint \vec{B} \cdot d\vec{l} = 2\pi\rho B_\rho = \mu_0\, i$$

Hence

$$\vec{B} = \frac{\mu_0 i}{2\pi\rho}\,\hat{\phi} \qquad ; |\vec{\rho}| < a$$

(c) In Prob. 4.6 we learned that there is an opposite charge induced on the inner surface of the cylinder to cancel the electric field inside the conductor. When seen from the moving frame, this charge *also* produces a current that is *equal and opposite* to the current from the moving line of charge. Now take the amperian loop in part (b) to lie inside the conducting cylinder. There is *no* net current flowing through the loop, and hence

$$\vec{B} = 0 \qquad ;\ a < |\vec{\rho}| < b$$

(d) Similarly, there is an opposite charge induced on the *outside* of the neutral conducting cylinder, which also gives rise to a current, so that when the amperian loop is enlarged to surround the conducting cylinder, one simply recovers the result in part (b)

$$\vec{B} = \frac{\mu_0 i}{2\pi\rho}\,\hat{\phi} \qquad ; b < |\vec{\rho}|$$

(e) From the frame moving with $-\vec{v}$, the electric field in Prob. 4.6 appears to be moving with velocity $+\vec{v}$, and Eq. (22.17) from special relativity states that a moving electric field has associated with it a magnetic field of the form

$$\vec{B} = \frac{1}{c^2}\vec{v} \times \vec{E}$$

In Prob. 4.6, the electric field lies in the $\hat{\rho}$-direction, and we observe that

$$\frac{\lambda}{\varepsilon_0 c^2}\vec{v} \times \hat{\rho} = \mu_0(\lambda v)\,\hat{v} \times \hat{\rho} = \mu_0\, i\,\hat{\phi}$$

The above results for the magnetic fields $\vec{B}$ are then reproduced from the electric fields $\vec{E}$ of Prob. 4.6!

What is the moral here? Just as in Prob. 22.5, we learn that electrostatics and special relativity can reproduce the results of Ampere's law and magnetostatics, at least in this case.

Problem 23.3 The magnetization in Probs. 15.1–15.3 can be related to the equivalent *surface current density* $\vec{j}_{\text{surface}}$ in the following fashion (Fig. 23.2 below).

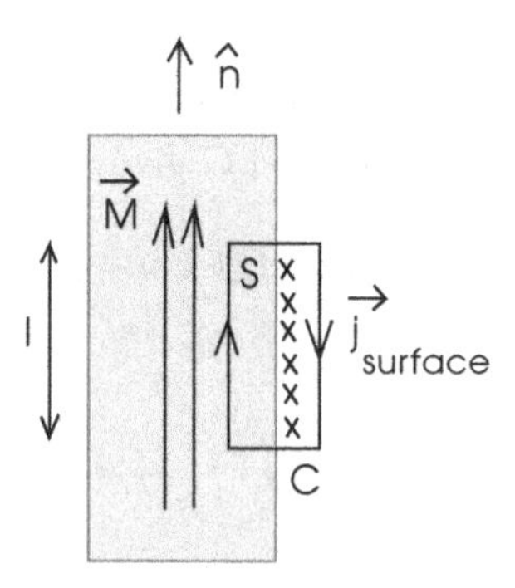

Fig. 23.2 Amperian loop used to demonstrate $\vec{M} = \mu_0 \eta_{\text{surface}}\, \hat{n}$. Both $\vec{j}_{\text{surface}}$ and $d\vec{S}$ point into the page.

Define the local quantity

$$\mu_0 \vec{j}_{\text{surface}} \equiv \vec{\nabla} \times \vec{M}$$

(a) Now consider the amperian loop in the figure. Use Stokes' theorem to show the surface current through the loop is given by

$$\mu_0 \int_S \vec{j}_{\text{surface}} \cdot d\vec{S} = \int_S \left(\vec{\nabla} \times \vec{M}\right) \cdot d\vec{S} = \oint_C \vec{M} \cdot d\vec{l}$$

(b) Show this reproduces the expression for $\vec{M}$ in Prob. 15.1

$$\vec{M} = \mu_0 \eta_{\text{surface}}\, \hat{n}$$

Solution to Problem 23.3

(a) Just evaluate

$$\mu_0 \int_S \vec{j}_{\text{surface}} \cdot d\vec{S} = \int_S \left(\vec{\nabla} \times \vec{M}\right) \cdot d\vec{S}$$

With the aid of Stokes' theorem for the final integral, this gives

$$\mu_0 \int_S \vec{j}_{\text{surface}} \cdot d\vec{S} = \int_S \left(\vec{\nabla} \times \vec{M}\right) \cdot d\vec{S} = \oint_C \vec{M} \cdot d\vec{l}$$

(b) The current flowing through the amperian loop is

$$I_{\text{surface}} = \int_S \vec{j}_{\text{surface}} \cdot d\vec{S} = \eta_{\text{surface}}\, l$$

where η_{surface} is the current per unit length.

From the figure, if the magnetization is written $\vec{M} = M_n\,\hat{n}$, then the evaluation of the loop integral around C gives

$$\oint_C \vec{M} \cdot d\vec{l} = M_n\, l + 0 + 0 + 0$$

Hence, from part (a),

$$\mu_0 \eta_{\text{surface}}\, l = M_n\, l$$

It follows that

$$\vec{M} = \mu_0 \eta_{\text{surface}}\, \hat{n}$$

Problem 23.4 Suppose one writes Maxwell's Eqs. (23.20) in a *medium* that has both the dielectric polarization $\vec{P}$ of Prob. 7.4 and the magnetization $\vec{M}$ of Prob. 23.3.

(a) Show the effect is to replace the sources by[3]

$$\rho \to \rho_{\text{free}} + \rho_{\text{pol}}$$
$$\vec{j} \to \vec{j}_{\text{free}} + \vec{j}_{\text{surface}} + \vec{j}_{\text{pol}}$$

where

$$\rho_{\text{pol}} \equiv -\varepsilon_0 \vec{\nabla} \cdot \vec{P}$$
$$\mu_0 \vec{j}_{\text{surface}} \equiv \vec{\nabla} \times \vec{M} \qquad ; \; \mu_0 \vec{j}_{\text{pol}} \equiv \mu_0 \varepsilon_0 \frac{\partial \vec{P}}{\partial t}$$

(b) Show that the continuity equation of Prob. 18.2 continues to hold;

(c) What are the appropriate expressions in the proper SI units of Prob. 15.5?

Solution to Problem 23.4

(a) In the presence of a dielectric, the applied field $\vec{D}$ arises from the free charge, so that the first Maxwell equation reads

$$\vec{\nabla} \cdot \vec{D} = \frac{1}{\varepsilon_0} \rho_{\text{free}}$$

[3]We suppress the $(\vec{x}, t)$ dependence.

From Prob. 7.4, the applied (displacement) field $\vec{D}$ is related to the electric field $\vec{E}$ and polarization $\vec{P}$ by

$$\vec{D} = \vec{E} + \vec{P}$$

Hence

$$\vec{\nabla} \cdot \vec{E} = \frac{1}{\varepsilon_0}\rho \qquad ; \rho \to \rho_{\rm free} + \rho_{\rm pol}$$

where the polarization charge density $\rho_{\rm pol}$ is given by

$$\rho_{\rm pol} \equiv -\varepsilon_0 \vec{\nabla} \cdot \vec{P}$$

In the presence of a magnetic material, the applied field $\vec{H}$ arises from the free current (and free charge), so that the fourth Maxwell equation reads

$$\vec{\nabla} \times \vec{H} = \mu_0 \vec{j}_{\rm free} + \mu_0\varepsilon_0 \frac{\partial \vec{D}}{\partial t}$$

From Prob. 15.1, the applied field $\vec{H}$ is related to the magnetic field $\vec{B}$ and magnetization $\vec{M}$ by

$$\vec{B} = \vec{H} + \vec{M}$$

Hence

$$\vec{\nabla} \times \vec{B} = \mu_0 \vec{j} + \mu_0\varepsilon_0 \frac{\partial \vec{E}}{\partial t} \qquad ; \vec{j} \to \vec{j}_{\rm free} + \vec{j}_{\rm surface} + \vec{j}_{\rm pol}$$

where the surface and polarization currents are given by

$$\mu_0 \vec{j}_{\rm surface} \equiv \vec{\nabla} \times \vec{M} \qquad ; \mu_0 \vec{j}_{\rm pol} \equiv \mu_0\varepsilon_0 \frac{\partial \vec{P}}{\partial t}$$

(b) Maxwell's Eqs. (23.20) now read

$$\begin{aligned}
\vec{\nabla} \cdot \vec{E}(\vec{x},t) &= \frac{1}{\varepsilon_0}\rho(\vec{x},t) \qquad ; \text{Maxwell's equations} \\
\vec{\nabla} \cdot \vec{B}(\vec{x},t) &= 0 \\
\vec{\nabla} \times \vec{E}(\vec{x},t) &= -\frac{\partial \vec{B}(\vec{x},t)}{\partial t} \\
\vec{\nabla} \times \vec{B}(\vec{x},t) &= \mu_0\, \vec{j}(\vec{x},t) + \mu_0\varepsilon_0\, \frac{\partial \vec{E}(\vec{x},t)}{\partial t}
\end{aligned}$$

where $(\rho, \vec{j})$ are given by the above. Take the divergence of the fourth equation, and make use of the first

$$
\begin{aligned}
0 &= \mu_0 \left[\vec{\nabla} \cdot \vec{j} + \varepsilon_0 \frac{\partial}{\partial t} \vec{\nabla} \cdot \vec{E} \right] \\
&= \mu_0 \left[\vec{\nabla} \cdot \vec{j} + \frac{\partial \rho}{\partial t} \right]
\end{aligned}
$$

Now substitute the above expressions for the current and charge densities in the presence of the medium[4]

$$
\mu_0 \left[\vec{\nabla} \cdot \vec{j}_{\rm free} + \varepsilon_0 \frac{\partial}{\partial t} \vec{\nabla} \cdot \vec{P} + \frac{\partial \rho_{\rm free}}{\partial t} - \varepsilon_0 \frac{\partial}{\partial t} \vec{\nabla} \cdot \vec{P} \right] = 0
$$

Hence, the continuity equation of Prob. 18.2 continues to hold

$$
\vec{\nabla} \cdot \vec{j}_{\rm free} + \frac{\partial \rho_{\rm free}}{\partial t} = 0
$$

(c) We can just apply the results in Prob. 15.5

$$
\begin{aligned}
&\rho_{\rm pol} \equiv -\vec{\nabla} \cdot \vec{\mathcal{P}} \\
&\vec{j}_{\rm surface} \equiv \vec{\nabla} \times \vec{\mathcal{M}} \qquad ; \; \vec{j}_{\rm pol} \equiv \frac{\partial \vec{\mathcal{P}}}{\partial t}
\end{aligned}
$$

In terms of the new auxiliary fields $(\mathcal{D}, \mathcal{H})$, the two Maxwell equations involving the sources now become

$$
\begin{aligned}
\vec{\nabla} \cdot \vec{\mathcal{D}} &= \rho_{\rm free} \\
\vec{\nabla} \times \vec{\mathcal{H}} &= \vec{j}_{\rm free} + \frac{\partial \vec{\mathcal{D}}}{\partial t}
\end{aligned}
$$

The simplicity of these results provides the primary motivation for taking out the constants (ε_0, μ_0) in Prob. 15.5.[5]

Problem 23.5 Given that electromagnetic radiation consists of *photons*, each with energy $E = h\nu$ and momentum $p = h\nu/c$, show that Eq. (23.25)

---

[4]Note that $\mu_0[\vec{\nabla} \cdot \vec{j}_{\rm surface}] = \vec{\nabla} \cdot (\vec{\nabla} \times \vec{M}) = 0$.

[5]It follows from Prob. 15.5 and Eqs. (7.18), (7.29), and (15.12) that in the medium

$$
\begin{aligned}
\vec{\mathcal{D}} &= \varepsilon_0 \kappa \vec{E} \\
\vec{B} &= \mu_0 \kappa_m \vec{\mathcal{H}}
\end{aligned}
$$

for the radiation pressure for absorption at normal incidence follows immediately

$$P_{\text{rad}} = \frac{1}{c}|\vec{S}\,| \qquad ; \text{ radiation pressure}$$

Solution to Problem 23.5

The energy flux (energy crossing unit transverse area per unit time) for electromagnetic radiation is given by the Poynting vector

$$\vec{S} = \frac{1}{\mu_0}\vec{E} \times \vec{B} \qquad ; \text{ energy flux}$$
$$\qquad ; \text{ Poynting vector}$$

The absorbed energy flux into a black body at normal incidence is then $|\vec{S}\,|$. As stated, electromagnetic radiation consists of *photons*, each with energy $E = h\nu$ and momentum $p = h\nu/c$. Every time a photon is absorbed in the energy transfer, an equivalent amount of momentum ($1/c$ times the energy) is thus also transferred. Hence, the radiation pressure for absorption at normal incidence is

$$P_{\text{rad}} = \frac{1}{c}|\vec{S}\,| \qquad ; \text{ radiation pressure}$$

****

(*Aside*) The frequency and wavelength of *real* photons satisfy $\nu\lambda = c$. The wavelength λ of *virtual* photons can be varied away from this. Electron scattering through the exchange of a virtual photon (Fig. 23.3 below), as carried out, for example, at the Continuous Electron Beam Accelerator Facility (CEBAF) at Jefferson Laboratory in Newport News, VA, is used to study the internal structure of nucleons and nuclei.[6]

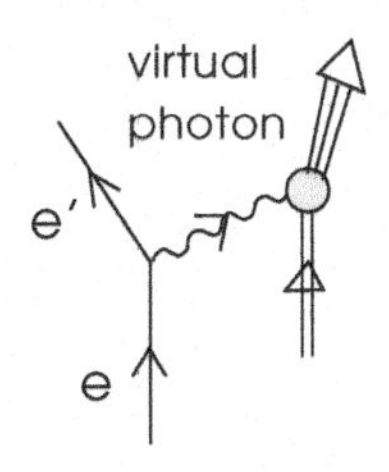

Fig. 23.3 Electron scattering through the exchange of a virtual photon.

****

[6]See [Walecka (2008); Walecka (2004)].

Problem 23.6 This is not a problem in the ordinary sense, but I just wanted to leave you with something to ponder. When one detects far red-shifted light coming from the most distant galaxies, one is observing electromagnetic radiation that has traveled through space, with the oscillating electric and magnetic fields described by Maxwell's equations, for almost the *entire age of the universe.* Good luck with your future courses. Enjoy!

Appendix A

Vector Identities

The following vector identities are proven in the solutions to these *Problems*:

$$(\vec{a} \times \vec{b}) \cdot (\vec{c} \times \vec{d}) = (\vec{a} \cdot \vec{c})(\vec{b} \cdot \vec{d}) - (\vec{a} \cdot \vec{d})(\vec{b} \cdot \vec{c})$$

$$\vec{a} \times (\vec{b} \times \vec{c}) = (\vec{a} \cdot \vec{c})\,\vec{b} - (\vec{a} \cdot \vec{b})\,\vec{c}$$

$$\vec{a} \cdot (\vec{b} \times \vec{c}) = \vec{c} \cdot (\vec{a} \times \vec{b}) = \vec{b} \cdot (\vec{c} \times \vec{a})$$

$$\vec{\nabla} \cdot (\vec{a} \times \vec{b}) = \vec{b} \cdot (\vec{\nabla} \times \vec{a}) - \vec{a} \cdot (\vec{\nabla} \times \vec{b})$$

$$\vec{\nabla} \cdot \left(\vec{\nabla} \times \vec{v}\right) = 0$$

$$\vec{\nabla} \times \left(\vec{\nabla}\chi\right) = 0$$

$$\vec{\nabla} \times \left[\vec{\nabla} \times \vec{v}\right] = \vec{\nabla}\,(\vec{\nabla} \cdot \vec{v}) - \nabla^2\,\vec{v}$$

Bibliography

Abraham, M., and Becker, R., (1949). *The Classical Theory of Electricity and Magnetism*, Hafner, New York, NY

Amore, P., and Walecka, J. D., (2013). *Introduction to Modern Physics: Solutions to Problems*, World Scientific Publishing Company, Singapore

Amore, P., and Walecka, J. D., (2014). *Topics in Modern Physics: Solutions to Problems*, World Scientific Publishing Company, Singapore

Amore, P., and Walecka, J. D., (2015). *Advanced Modern Physics: Solutions to Problems*, World Scientific Publishing Company, Singapore

Fetter, A. L. and Walecka, J. D., (2003). *Theoretical Mechanics of Particles and Continua*, McGraw-Hill, New York (1980); reissued by Dover Publications, Mineola, New York

Fetter, A. L. and Walecka, J. D., (2003a). *Quantum Theory of Many-Particle Systems*, McGraw-Hill, New York (1971); reissued by Dover Publications, Mineola, New York

Fetter, A. L., and Walecka, J. D., (2006). *Nonlinear Mechanics: A Supplement to Theoretical Mechanics of Particles and Continua*, Dover Publications, Mineola, New York

Freedman, R., Ruskell, T., Keston, P. M., and Tauck, D. L., (2013). *College Physics*, W. H. Freeman, San Francisco, CA

Griffiths, D. J., (2017) *Introduction to Electrodynamics, 4th ed.*, Cambridge U. Press, Cambridge, UK

Halliday, D., Resnick, R., and Walker, J., (2013). *Fundamentals of Physics, 10th ed.*, J. Wiley and Sons, New York, NY

Jackson, J. D., (2009). *Classical Electrodynamics, 3rd ed.*, student ed., J. Wiley and Sons, New York, NY

Ohanian, H. C., (1985). *Physics*, W. W. Norton and Co., New York, NY

Ohanian, H. C., (1995). *Modern Physics, 2nd ed.*, Prentice-Hall, Upper Saddle River, NJ

Panofsky, W. K. H., and Phillips, M., (2005). *Classical Electricity and Magnetism*,

Dover Publications, Mineola, NY

Purcell, E. M., and Morin, D. J., (2013). *Electricity and Magnetism, 3rd ed.*, Cambridge U. Press, Cambridge, UK

Schwarz, M., (1987). *Principles of Electrodynamics*, Dover Publications, Mineola, NY

Slater, J. C., and Frank, N. H., (2011). *Electromagnetism*, Dover Publications, Mineola, NY

Stratton, J. A., (2008) *Electromagnetic Theory*, Adams Press, *adamspress.com*

Walecka, J. D., (2000). *Fundamentals of Statistical Mechanics: Manuscript and Notes of Felix Bloch, prepared by J. D. Walecka*, World Scientific Publishing Company, Singapore; originally published by Stanford University Press, Stanford, CA (1989)

Walecka, J. D., (2004). *Theoretical Nuclear and Subnuclear Physics, 2nd ed.*, World Scientific, Singapore

Walecka, J. D., (2007). *Introduction to General Relativity*, World Scientific Publishing Company, Singapore

Walecka, J. D., (2008). *Introduction to Modern Physics: Theoretical Foundations*, World Scientific Publishing Company, Singapore

Walecka, J. D., (2010). *Advanced Modern Physics: Theoretical Foundations*, World Scientific Publishing Company, Singapore

Walecka, J. D., (2011). *Introduction to Statistical Mechanics*, World Scientific Publishing Company, Singapore

Walecka, J. D., (2013). *Topics in Modern Physics: Theoretical Foundations*, World Scientific Publishing Company, Singapore

Walecka, J. D., (2017). *Introduction to Statistical Mechanics: Solutions to Problems*, World Scientific Publishing Company, Singapore

Walecka, J. D., (2017a). *Introduction to General Relativity: Solutions to Problems*, World Scientific Publishing Company, Singapore

Walecka, J. D., (2018). *Introduction to Electricity and Magnetism*, World Scientific Publishing Company, Singapore

Wiki (2017). *The Wikipedia*, http://en.wikipedia.org/wiki/(topic)

Index

Printed in the United States
By Bookmasters